Research Within Reach

Elementary School Mathematics

Mark J. Driscoll

CEMREL, Inc.
National Institute of Education

 NATIONAL COUNCIL OF TEACHERS OF MATHEMATICS
1906 Association Drive, Reston, Virginia 22091

This book brings under one cover a series of bulletins on elementary school mathematics which were originally published as separate pieces by the Research and Development Interpretation Service (RDIS) for the Research and Development Exchange (RDx).

Table of Contents

Mathematical Development: The Children
    Mathematics in Kindergarten   7
    The Bridge from Concrete to Abstract   11
    Counting Strategies   17
    The Role of Manipulatives in Elementary School Mathematics   21
    Measurement in Elementary School Mathematics   29

Mathematical Development: The Teacher's Role
    Diagnosis: Taking the Mathematical Pulse   39
    Unlocking the Mind of a Child:
    Teaching for Remediation in Mathematics   45
    Evaluation in Mathematics Education Part One:
    Looking Beneath and Beyond the Tests   51
    Evaluation in Mathematics Education Part Two:
    Mastery Learning in Elementary School Mathematics   56
    Motivation in Mathematics   61

Instructional Strategies
    Meaning in Elementary School Mathematics   67
    Securing Mathematical Skills: Drill and Other Topics   72
    Grouping for Elementary School Mathematics   79
    Learning Elementary School Mathematics:
    Individual Styles and Individual Needs   84
    Algorithms in Elementary School Mathematics   91

Stronger Curriculum
    Mathematical Problem Solving: Not Just a Matter of Words   101
    Estimation and Mental Arithmetic   107
    Calculators in the Classroom   112
    Sequence with Substance:
    The Elementary School Mathematics Curriculum   116
    The Teacher and the Textbook   123

Indexes
    Subject Index   131
    Author Index   135

Research
Within
Reach
Elementary School Mathematics

Introduction

Elementary school teachers and curriculum specialists are faced daily with the need to make complex instructional decisions. The means teachers use for solving their problems are varied. Frequently, they review the teacher's guides that accompany mathematics textbooks series or sets of supplementary materials, consult the curriculum guides developed and provided by the school district, and discuss their concerns or problems with co-workers. Occasionally, they consult their supervisory and support staffs.

The one source of information most teachers do not consult is the research literature—they feel that research activity is so far removed from the classroom that it will not help them solve their immediate instructional problems. And yet elementary school mathematics has been, and continues to be, a fertile area for research activity.

Is the application of research to instructional practice important in the field of mathematics? Can the results of research be used to improve teaching? The answer to each of these questions is "yes," but if this is true, why is it that practitioners seem to automatically discount the usefulness of research literature? We can discover the reasons by looking at the nature of research and how research results are disseminated.

First, researchers traditionally have been reluctant to speculate about the educational implications of their work. Instead, the work they do often results in the need for further research. These extensions of the work typically take a closer, narrower look at the construct under investigation; thus, with successive refinements of the research, the results tend to get less and less generalizable, and certainly less practical.

Second, because the jargon, methodologies, and concerns of researchers and practitioners are usually so different, there has been an inadequate exchange of knowledge between the groups. The usual form for exchanging scholarly information is through journals and paper presentations at professional meetings. However, most practitioners do not read the *Journal of Research in Mathematics Education* or attend the annual meeting of the Research Council for Diagnostic and Prescriptive Mathematics. Whatever usefulness could be made of basic research cannot be made unless it is received by practitioners.

There have, of course, been a number of efforts to synthesize and interpret research findings in the area of mathematics instruction. Most of these synthesis products—and some of them are highly respected—began with a review of the literature and moved from there to discuss classroom implications. Research and Development Interpretation Service (RDIS) project staff were not satisfied with that approach, as you will see.

A New Approach

We were not interested in simply reporting the results of a review of the research literature to mathematics educators. A group of people who have been thinking and writing for some time about the processes involved in knowledge transformation worked hard to conceptualize a process that would ensure the relevancy of the work we would be doing. Out of these discussions came the decision to actually go to elementary educators and ask them to describe for us their most pressing concerns and needs in relation to mathematics instruction.

RDIS obtained a list of the names of 2000 elementary school teachers drawn from all 50 states and distributed evenly from kindergarten through grade six. A total of 350 teachers were randomly selected from the list and were asked by letter to participate in a 30-minute telephone interview. Of those, 38 responded favorably, and each engaged in an interview with an RDIS staff member, probing the issues they felt were most pressing in day-to-day mathematics teaching. As they uncovered their concerns, the teachers were guided to translate them into questions directed to the research community.

The next step in our approach to increasing the communication between researchers and practitioners was the selection of an advisory consultant panel of nationally prominent researchers. We asked representatives of the mathematics research community to nominate ten authorities in the field of mathematics whom they would recommend to assist us in the task of providing research-guided answers to practitioners' questions. The individuals named most often were asked to serve on our consultant panel.

The members of the panel are Robert E. Reys, University of Missouri at Columbia, Marilyn N. Suydam, Ohio State University and the ERIC Clearinghouse for Science, Mathematics and Environmental Education, and James W. Wilson, University of Georgia, Athens.

The consultant panel met initially with RDIS project staff to discuss the practitioner questions and decide which questions were most amenable to research-guided responses. The members of the panel felt strongly that the final product of their efforts should be a series of "research bulletins" rather than a monograph. These bulletins would deal with a total of 21 topics and would appear as the drafts were completed. The draft of each bulletin that has resulted from these early discussions has gone through an extensive review process, including reviews at several stages by the consultant panel and by several additional mathematics educators—both researchers and practitioners. Several of the practitioners who were interviewed have been asked for their criticisms and suggestions.

Research Within Reach: Elementary School Mathematics is a series of bulletins addressing questions that arise out of actual classroom situations. Each bulletin begins with a question which sets the stage for the discussion that follows. *Research Within Reach* reflects research conclusions, wherever it is possible to draw conclusions. To the extent possible, the consultant panel members have based their recommendations on sound evidence. They have also relied upon their personal opinions, experiences and common sense. Their purpose has not been to give the final analysis on any of the topics addressed. Research is being conducted now that will require that we take a close look at them again in the near future.

We want to emphasize the importance of the suggested readings at the end of each bulletin. The references were selected for their diversity. Some are quite technical, and others are more narrative. Some are literature review papers, some are papers of opinion, and some are reports of research. Some reflect the predominant position of researchers and other professionals in the field (which are presented throughout the bulletins), whereas others represent an opposing point of view.

To further guide the practitioner, the readings in each selection that would be especially interesting to and useful for teachers are highlighted. These references are preceded by an asterisk (*). Additional readings can be chosen according to special needs or areas of interest.

A Final Note

The hope of RDIS staff is that this series of bulletins will prove useful to practitioners. If you find it thought-provoking or have suggestions concerning the content of the document or the process used, please let us know. We need your input if we are to succeed in our goal of increasing communication between the research and development community and educational practitioners.

Mathematical Development: The Children

Research Within Reach
Elementary School Mathematics

Mathematics in Kindergarten

Question: I teach kindergarten and am frustrated by the way the mathematics in kindergarten textbooks is organized. It jumps around from number ideas to shape to color concepts. Where should the areas of content emphasis be for teachers of kindergarten mathematics?

Already little mathematicians when they begin formal schooling, kindergarteners move through their world with a wealth of practical (not yet formal or school-initiated) arithmetic skills.

At the same time, we know that these children are naturally limited, by their level of cognitive development, in the kinds of mathematical experiences they can meaningfully undertake.

It is imperative—if we are not to shortchange or confuse young children—that we look carefully at what they bring by way of mathematical intuition and experience when they begin school, and then look at what we can offer them in formal training and curriculum. Will it allow them to continue to develop mathematically as they have already begun to develop on their own?

Established Mathematical Skills

Most children who are beginning kindergarten deal comfortably with situations which require perceptions of "largest," "smallest," "tallest," "longest," "inside," "beside," "most," "closest," and "farthest" (11).

Their perceptions of size difference have already begun to operate in the conceptual realms of addition and proportion. For example, Brush (1) and Ginsburg (5), found that a number of children between the ages of four and six can "add," in a mature way, in the following setting:

> Jar A and jar B have marbles in them. After the child decides the difference, if any, in size of the marble collections ("A has more marbles."), the jars are removed from sight and the child watches as a few more marbles are added to A (still out of sight). The child then responds to the question, "Which jar has more marbles?"

The researchers found that over 80 percent of the children were able to keep in mind the original equality or inequality, to consider the amount added, and then to make the appropriate decision about the final comparison. The numbers used were relatively small (15 or less) but there was strong evidence that the children were making their comparisons without counting (1, 5).

Similar experiments have revealed that four- to six-year-old children can make proportion estimates in quick fashion when they are shown pairs of collections of objects. They are able, for example, to indicate that a collection of ten marbles is closer in size to a collection of twelve marbles than a collection of twelve is close to twenty. Again, they seem to do this without reference to number or counting (5).

Even without much explicit understanding of number or counting, then, entering kindergarteners bring with them the solid intuitive groundwork for learning addition, proportion, and subtraction (inherent in their perceptions of size difference).

- Many of these children, however, do bring to school skills related to counting and number. One study of children entering kindergarten found that 75 percent could correctly rote count to ten or beyond (that is, counting recitation without the counting of objects), and that 50 percent could rote count to fourteen or beyond (10). Other research has revealed that many four- or five-year-old children recognize sets of one, two, three, or four objects and can recognize objects cut into halves or thirds (11).

Far from offering a mathematically blank slate, the beginning kindergartener offers real and practical, though roughhewn, arithmetic skills. A wise teacher will make use of these skills in guiding the children into the formal learning of mathematics.

What Teachers Can Do

The child's level of cognitive development dictates some clear mathematical limitations. For example, four or five year olds cannot verbalize at a level consistent with their use of many concepts. Furthermore, they tend to be very susceptible to distractions by irrelevant detail. As one result, they usually fall short of "number conservation," the ability to discern, for example, that the number of marbles in a row doesn't increase just because the row is spread out or doesn't diminish just because the row is shortened. Until they can conserve numbers, children are not ready to move very far into formal arithmetic (12).

However, there are numerous mathematical activities that kindergarten teachers can introduce to enrich the move *toward* formal arithmetic.

- For one thing, they can capitalize on and help develop the child's burgeoning ability to compare sizes and quantities. Young children are intimately concerned with their own height. Consequently, height comparison between children, between objects standing, between objects lying down (really length comparisons), and so on, present themselves as natural development exercises. (The 37th Yearbook of the National Council of Teachers of Mathematics, *Mathematics Learning in Early Childhood*, has many such examples; see especially "Experiences for Young Children" by Gibb and Castaneda). However, the teacher who engages his or her class in such activities should be working to develop in the children the specific language that is required for describing particular comparisons. ("How do you know that Billy is taller than Mike?")

- From *comparing* pairs of objects, children can move to *ordering* similar objects according to acknowledged rules—longest to shortest or lightest (in weight) to heaviest are among the possibilities.

- While they are being regularly exposed to phrases like "belong together" and "are alike," children can work on *classifying*. For example, the teacher can ask a child to approach a box of varied objects and find two objects that are alike. Again, the child's learning to verbalize should be a constant teacher goal, so it would be wise to follow the child's choice by asking her to tell how the objects are alike.

Because of the possibility that young children will consider an object as belonging inherently to one class and to no other, it is important for teachers to push for some flexibility in classifying, providing exercises in which the children can see, for example, that a red ball can be in a collection of red things one minute and in a collection of round things soon afterward.

- The work done with classifying, comparing, and ordering must be done without reference to numbers before the children can be expected to do the same with numbers. As a bridge to number work, teachers can offer experience in making *one-to-one correspondence* between two sets, or *matching*. For example: The teacher can give one child three jars of paste and another child three paste brushes, asking, "Is there a brush for each jar of paste? A jar for each paste brush? How can you find out?" With some consistent work like this, the child should move smoothly into comfortable use of the phrase "as many as."

- When the children become practiced in matching objects, they can begin to count them out loud to see how many are in a particular set. (They should still be encouraged to make physical matchings of the objects.) As the year progresses, kindergarten teachers can reasonably expect most of their students to be able to put in order the numbers from 0 to 10 (9). In other words, if sets of 1, 5, and 9 marbles are laid out on a table, the child should be aware by the end of the year that a set of 3 marbles would be placed between the set of 1 and the set of 5.

Looking Ahead

Kindergarten serves as a bridge to more formal schooling, and kindergarten teachers should always have an eye on the future mathematical needs of their children.

- One of those future needs is the learning of place value. As a reasonable goal in preparation for place value, teachers can expect that most of their children will leave kindergarten able to separate sets of up to ten items into subsets and able to name the number of items in each subset (9).

For reinforcing the learning in this area, as well as for introducing grouping by tens, teachers can look to several effective models. Beansticks, for example, can provide place value groundwork while they reinforce counting skills and help children organize information. (Beansticks are made with dried beans, tongue depressors— or popsicle sticks—and glue. For work on place value, single beans can represent units, a stick onto which ten beans have been glued can represent a ten, and a raft of ten-sticks can represent a hundred. See Merseth [7] and Wirtz [13] for further guidance.)

- Over and above preparation for specific topics like place value, teachers can help children develop some familiarity with skills that will sustain them as mathematics becomes more and more abstract to them. In particular, young children need to be exposed to patterns and regularity in mathematics and to ways of organizing and recording mathematical information. For the former, see Dawes (3), Ginsburg (5), and McKillip (6); they offer some rich examples, such as the following:

Can you continue this pattern, using all the red and white blocks you have?

For suggestions on organizing and recording, see Carpenter, Hiebert, and Moser (2) and McKillip (6). For example, even very young children are capable of moving from standing in groups according to month of birthday, to putting name labels inside hoops that are labelled with appropriate birthday months, even to understanding a month-by-month birthday graph. (On graph paper with large squares, each child can color in a square in the column beneath his or her birthday month.) Such graphing experiences not only help children in organizing information but also open the way for problem solving and number experiences ("In which month does the class have the most birthdays?").

Conclusion

One deliberate focus of this bulletin has been a look at the marvelous ways kindergarten children can develop their mathematical skills by practical, informal, nonwritten routes. However, this in no way is intended to imply that an introduction to formal arithmetic should be rushed. On the contrary, some recent research reveals the enormous potential for beginning first graders to succeed in tackling addition and subtraction *word problems* before they have had any training in formal addition or subtraction (2). The message is clear for kindergarten teachers (and for primary teachers): Be a guide; provide opportunities. Use what the children bring to you before you take them into unfamiliar territory.

References

1. Brush, L.R., and Ginsburg, H. "Preschool Children's Understanding of Addition and Subtraction." Unpublished manuscript, Cornell University, 1971.
2. Carpenter, Thomas P., James Hiebert, and James Moser. "The Effect of Problem Structure on First-Graders' Initial Solution Processes for Simple Addition and Subtraction Problems." Madison, Wisconsin: Wisconsin R&D Center, 1979.
* 3. Dawes, Cynthia. *Early Maths*. New York: Longman Group Limited, 1977.
4. Gibb, E. Glenadine, and Alberta M. Castaneda. "Experiences for Young Children." In J.N. Payne (ed.), *Mathematics Learning in Early Childhood*, 37th NCTM Yearbook. Reston, Virginia: National Council of Teachers of Mathematics (NCTM), 1975.
5. Ginsburg, Herbert. "Young Children's Informal Knowledge of Mathematics." *The Journal of Children's Mathematical Behavior*, 1 (Summer 1975): 63-157.
6. McKillip, William D. "'Patterns'—A Mathematics Unit for Three and Four Year Olds." *The Arithmetic Teacher*, 17 (January 1970): 15-18.
* 7. Merseth, Katherine Klippert. "Using Materials and Activities in Teaching Addition and Subtraction Algorithms." In M. Suydam and R. Reys (eds.), *Developing Computational Skills*, 1978 NCTM Yearbook. Reston, Virginia: NCTM, 1978.
* 8. Nelson, L. Doyal, and Joan Kirkpatrick. "Problem Solving." In J.N. Payne (ed.), *Mathematics Learning in Early Childhood*, 37th NCTM Yearbook. Reston, Virginia: NCTM, 1975.
* 9. Payne, Joseph N., and Edward C. Rathmell. "Number and Numeration." In J.N. Payne (ed.), *Mathematics Learning in Early Childhood*, 37th NCTM Yearbook. Reston, Virginia: NCTM, 1975.
10. Rea, Robert E., and Robert E. Reys. "Mathematical Competencies of Entering Kindergarteners." *The Arithmetic Teacher*, 17 (January 1970): 65-74.
11. Riedesel, C.A. "Recent Research Contributions to Elementary School Mathematics." *The Arithmetic Teacher*, 17 (March 1970): 245-252.
*12. Suydam, Marilyn N., and J. Fred Weaver. "Research on Mathematical Learning." In J.N. Payne (ed.), *Mathematics Learning in Early Childhood*, 37th NCTM Yearbook. Reston, Virginia: NCTM, 1975.
13. Wirtz, Robert. *Drill and Practice at the Problem Solving Level*. Washington, D.C.: Curriculum Development Associates, 1974.

*These are the readings we believe would be especially interesting to and useful for teachers.

Research Within Reach
Elementary School Mathematics

The Bridge from Concrete to Abstract

Question: In my third-grade class every year there are a few students who have no sense of numbers at all. For example, they have no idea of what a "54" is. What can I do to help them get an abstract sense of number?

For the child between the age of four years and adolescence, the passage from concrete to abstract is not so much a bridge as it is the entire journey. A level of abstraction can be reached by age thirteen that was only a promise at age four. In the intervening years, the process of abstracting has grown through several successive levels.

To use an example from Lovell, a four-year-old child, walking through a garden, will say "snail" every time he sees a snail. However, he cannot be sure if he is seeing the same snail each time, or if he has seen several individuals, all of whom share a common "snailness" (4). In similar fashion, if he has three snails and three jars, he cannot be sure he will succeed at putting one snail in each jar until he actually places them there. He can't abstract the threeness from the two sets with the same assurance that will be his when he is older.

Levels of Abstraction

"Abstract" can be defined with various degrees of technicality. In this bulletin, we offer as a reference point Lovell's informal definition, "at all levels, abstraction indicates a way of organizing objects, events, qualities, and so forth" (4).

What, then, keeps the young child from giving her world the kind of mental organization she'll have at hand when she's older? Research and experience tell us that she is bound by her early cognitive growth and has not yet acquired some of the cognitive skills generally developed during adolescence. For example, until they are about seven or eight (for some children it is a year or two earlier, for others a year or two later) children are easily swayed by distractions as their minds organize perceptions. In problem solving situations, such factors as background color, the shape of a glass container, or the density of colored dots arranged in a picture for number recognition can distract a young child away from correct problem solutions.

Secondly, until they are seven or eight, children cannot usually hold more than one idea or variable in mind at any one time. Hence, if an adult changes both the width and the length of a moldable clay cylinder without changing the amount of clay, a young child cannot handle the simultaneous changes of width and length. The child is likely to conclude that the resulting cylinder has either more or less clay in it. (In a similar though nonmathematical vein, a parent who asks an angry or disappointed youngster to try and look at a particular situation through the parent's eyes may well be asking for the impossible.)

Thirdly, during those years before seven or eight—called the "preoperative" stage by Piaget—children find it difficult, if not impossible, to look back at a particular thought process and to be aware of contradictions in logical decisions they have made.

Other cognitive stumbling blocks for young children which have shown up in research studies are described by Stevenson (8). In essence, the preoperative child finds it difficult to monitor her own thoughts and to distinguish between what's "out there" (her perceptions) and the order and organization which her mind gives to those perceptions (her abstractions).

From about seven until about twelve or thirteen (the "concrete operations" stage described by Piaget), the child deals more effectively with distractions, multiple ideas, and contradictions than he was able to when he was younger. He is able to classify objects, count objects with an appreciation of number, and order objects according to size, quantity, and set inclusion ("all robins are birds, but not all birds are robins"). In fact, children during this stage develop many of the abstracting skills they will have as adults.

There are some skills, however, that are not developed until they reach adolescence and yet another level of abstraction. It is at this later level that children can take one more step away from their dependency on perceptions as catalysts for abstracting and are able to draw abstractions from other abstractions.

For example, although four year olds have some primitive intuitions of proportionality (see our *Research Within Reach* bulletin "Mathematics in Kindergarten"), it is apparently not until they are adolescents that they can fully engage in the complex process of recognizing the ratio relationships of 3 to 12 and of 5 to 20, and of then moving beyond that to a recognition that the two pairs of numbers stand in the same proportion (4). In such a process a new concept arises from the bonding of two other concepts, the hallmark of this second level of abstraction (Piaget's "formal operations" stage).

The Teacher's Role

In any elementary school classroom, there are likely to be children at several different levels of cognitive development—some who think of 54 merely as the immediate neighbor of 53 and 55 in a counting sequence and others who are aware of 54 as 5 tens, 4 ones; some who rely on memory alone for use of the subtraction algorithm and others who attach some mathematical meaning to the regrouping in

$$\begin{array}{r} {}^5\!\!\!/\!\!\!6\,{}^1\!\!\!/\!\!\!4 \\ -25 \\ \hline 39 \end{array}$$

Once they have assessed the levels of development in their classes, teachers are faced with the realization that different student needs will, to some extent, invite different teacher strategies. As an organizational strategy, teachers might elect to group the children according to developmental level. We make but a passing reference to this option and encourage you to read the *Research Within Reach* bulletins "Diagnosis: Taking the Mathematical Pulse" and "Grouping for Mathematics."

No matter what form classroom organization takes, there are suggestions drawn from research and established practice that can help to illuminate the teacher's role as the child progresses from concrete to abstract thinking.

The Teacher's Role: Manipulatives

- First of all, it is important that *all* elementary teachers stay mindful of the importance of manipulatives and pictorial representations in this growth process. Even at the fifth- and sixth-grade levels, many children's abstractions are still very much anchored in their perceptions of the world (4). Therefore, the opportunity to see concretely or pictorially that 1400 is 10 times as large as 140, or that 2/3 of 2/3 is 4/9, is vital to their development.

Furthermore, it is critical not only that teachers be mindful of different representations but that they consistently make explicit the linking of steps in the progression toward the symbolic representation of any mathematical concept.

An example can be found in the question introducing this bulletin—what can a teacher do when he or she suspects that children have been slow to develop a notion of numeration—that is, when they have a weak conceptual grasp of numbers which are greater than 9, such as 24? As Payne and Rathmell point out (6), the teacher should first check some prerequisites, namely

1. that each of the children is able to recognize the numbers from 0 to 9 as they are represented by concrete or pictorial means; and
2. that each child is able to partition any number between 2 and 9—that is, is able to show with blocks, dots, etc., that a group of 8 is a group of 5 together with a group of 3.

Having noted these two prerequisites, Payne and Rathmell present a sequence of units that guides a child from grouping objects (e.g., sticks or rods) into equivalent sets (e.g., tens) and naming the number of sets ("I have 2 groups of 10 sticks and 4 sticks left over"), to a scheme for understanding that one can group more than once (e.g., 10 tens can be grouped together as 100), to a scheme for recording the groups, to the final step of representing a number, say 24, in three ways—as 2 tens, 4 ones

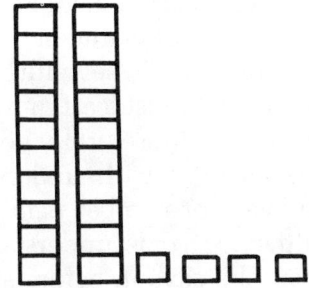

as the oral "twenty-four"; and as the written "24."

For a similar sequence directed at the addition and subtraction algorithms, see Merseth (5). Also read the *Research Within Reach* bulletin "Manipulatives in Elementary School Mathematics."

- One final note on the role of manipulatives in the bridge from concrete to abstract: research implies that students can benefit from exposure to multiple representations of the same concept (for example, $5 \times 12 = 60$ can be represented with base-ten blocks, by the use of grids on a piece of graph paper, and so on). However, the mathematical commonalities should be pointed out by the teacher if the students themselves do not recognize them (9).

The Teacher's Role: Language

- As provider of the fabric for conceptual organization and communication, language is closely tied to abstraction. It is natural to wonder if abstraction is the product of language learning and how much precise verbalization of mathematical ideas should be stressed in the classroom. The research shows that young children can understand and use mathematical concepts (such as "roundness") without knowing or using the words (8). As we noted before, the children in one kindergarten study were even able to indicate nonverbally an intuitive, though primitive, understanding of proportion (3). On the other hand, some children show a facility for mathematical language before they fully grasp the ideas behind the words. Teachers, particularly at the primary level, should not judge a child's conceptual understanding merely from the child's ability to verbalize.

- It is important that teachers regularly expose children to the oral use of mathematical terms, for example, "as many as," "seventh," and "in order of size." Children must learn how to use a mathematical word or phrase orally before they are expected to write it with understanding. This holds true even for number names, where the child should have some understanding of "fifty-four" when she says it, before she is expected to write "54" (6).

- In the area of curriculum development, at least one program, the Comprehensive School Mathematics Program (CSMP), has been developed with one eye toward circumventing the language barrier for young children. The program has developed three of its own "languages" to help children more readily approach mathematical problems (2). For example, the following first-grade problem is in the language of arrows:

The solid arrows stand for +3.
The broken arrows stand for 2×.
Label all the dots.

The Teacher's Role: Perceptions

- The growing body of research on brain hemispheric specialization indicates that young children are predisposed to spatial modes of learning (10). Their spatial perceptions, especially their *visual* perceptions, strongly influence their mathematical conceptualization. Thus, it is important that teachers make use of this orientation as they guide children from concrete to abstract thinking.

For example, an important step along the way to abstract mathematics is an awareness of the regularity occurring throughout mathematics (the order in which numbers are added doesn't make any difference—6 + 5 = 5 + 6; multiples of 10 all end in 0). Teachers should therefore take pains that children actually *see* mathematical regularity as much as possible—on the blackboard, with concrete objects, or using hand-held calculators.

- Furthermore, since children are so reliant on their perceptions, teachers should help them form mathematical generalizations, making sure they do not associate "triangle" only with right triangles. A good development of a mathematical concept like "triangle" includes many examples. It also should provide non-examples to enhance the concept formation (7). These experiences should provide a springboard to a discussion of commonalities and then build upon recognized commonalities and differences to develop better conceptual understanding.

Conclusion

In reviewing the decades of research on human growth in abstraction, Callahan and Glennon were moved to say that there are no final answers, that perhaps there is no single route to abstraction (1). We know that the rate at which children pass from concrete to abstract levels of thinking differs from child to child; perhaps, too, the routes themselves differ. Whatever the route or rate, teachers can still make the passage smooth. This report has been a presentation of some of the important factors in this smoothing process.

References

* 1. Callahan, Leroy G., and Vincent J. Glennon. *Elementary School Mathematics: A Guide to Current Research.* Washington, D.C.: Association for Supervision and Curriculum Development, 1975.

2. *CSMP in Action.* St. Louis, Missouri: CEMREL, Inc., Comprehensive School Mathematics Program, 1978.

3. Ginsburg, Herbert. "Young Children's Informal Knowledge of Mathematics." *The Journal of*

References (continued)

Children's Mathematical Behavior, 1 (Summer 1975): 63-157.

4. Lovell, Kenneth. *The Growth of Understanding in Mathematics: Kindergarten through Grade Three*. New York: Holt, Rinehart and Winston, 1971.

* 5. Merseth, Katherine Klippert. "Using Materials and Activities in Teaching Addition and Subtraction Algorithms." In M. Suydam and R. Reys (eds.), *Developing Computational Skills*, 1978 NCTM Yearbook. Reston, Virginia: National Council of Teachers of Mathematics (NCTM), 1978.

6. Payne, Joseph N., and Edward C. Rathmell. "Number and Numeration." In J.N. Payne (ed.), *Mathematics Learning in Early Childhood*, 37th NCTM Yearbook. Reston, Virginia: NCTM, 1975.

7. Shumway, Richard. "Students Should See 'Wrong' Examples: An Idea from Research on Learning." *The Arithmetic Teacher*, 21 (April 1974): 344-348.

8. Stevenson, Harold W. "Learning and Cognition." In J.N. Payne (ed.), *Mathematics Learning in Early Childhood*, 37th NCTM Yearbook. Reston, Virginia: NCTM, 1975.

* 9. Suydam, Marilyn N., and Jon L. Higgins. *Activity-Based Learning in Elementary School Mathematics: Recommendations from Research*. Reston, Virginia: NCTM, 1977.

10. Wheatley, Grayson H., et al. "Hemispheric Specialization and Cognitive Development: Implications for Mathematics Education." *Journal for Research in Mathematics Education*, 9 (January 1978): 19-32.

*These are the readings we believe would be especially interesting to and useful for teachers.

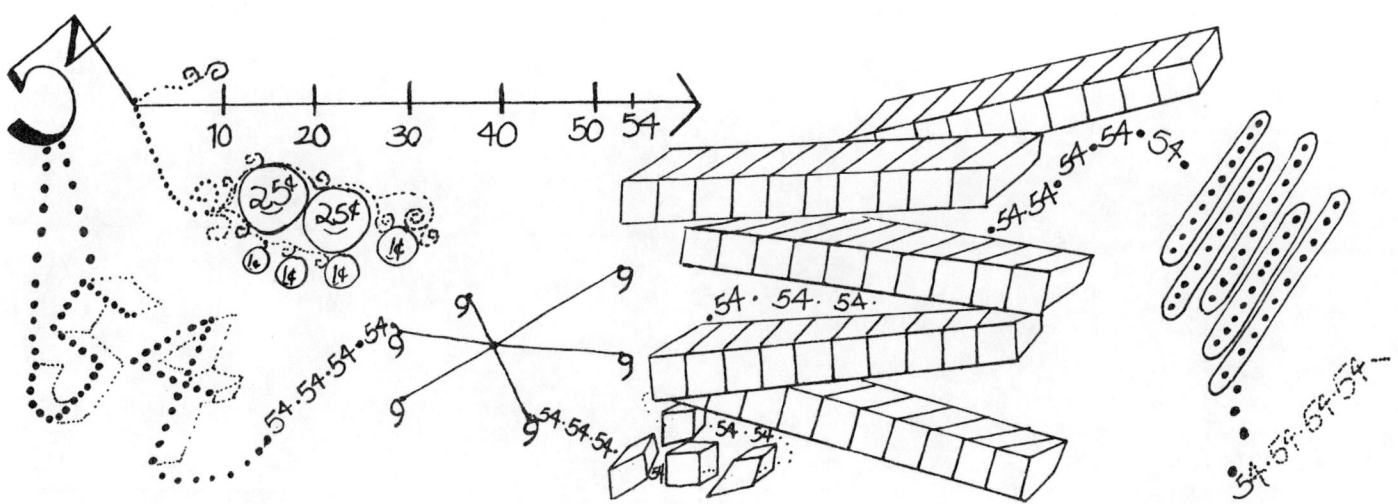

Research Within Reach
Elementary School Mathematics

Counting Strategies

Question: My first graders rarely count from starting points other than one. Is it important for me to give them supplemental work in counting-on? If so, how can it be done?

Counting, as a child behavior, can sometimes seem a mixed blessing. It is a source of pride for parents when it first appears and later becomes the springboard for learning concepts like addition and subtraction. Unfortunately, many students rely so heavily on counting—finger-counting, in particular—that they fail to adopt more efficient ways to do mathematics.

One recent study of fourth graders' strategies for column addition revealed that *nearly half* of the group (an above average group by national testing standards) resorted to finger-counting on a regular basis. Not surprisingly, the students using finger-counting were slower to complete addition exercises than those who relied only on their knowledge of basic facts. They also attempted fewer problems than those who didn't count on their fingers, and they had significantly fewer correct answers (13).

Because children have a need for efficient computational methods as mathematics becomes more and more sophisticated for them, teachers are well-advised to be concerned when counting seems too dominant.

On the other hand, counting strategies are vital to children's early learning of mathematics. They allow children to take hold of a variety of concepts, and they form the groundwork for later, more sophisticated thinking strategies. In fact, several studies (2, 5) have confirmed that incoming first graders, with no previous formal instruction in addition or subtraction, can solve a variety of word problems that involve addition, subtraction, and even some multiplication and division. Counting is a valuable and much used tool for young children and strongly influences their problem solving. The study by Carpenter and his colleagues (2) found that counting strategies abound among young children. They also found that, using their counting strategies, very few children apply the wrong operation in their solutions, such as adding when subtraction is appropriate.

17

Since this error has been observed primarily with older children who have already experienced formal instruction in addition and subtraction, it may be a result of learning symbolic representations. In typical classroom procedure, addition and subtraction are introduced in terms of joining or separating sets using either pictures or concrete objects. Then children are drilled on abstract problems with number sentences. When they finally get to verbal problems, their response is, "Is this a plus or a takeaway?" In this format the operations are initially learned outside the context of verbal problems. When verbal problems are introduced later, children are simply told that addition and subtraction can be used to solve these problems, but they have no basis for using their natural intuition to relate the problem structure to the operations they have learned. ... This may result not only in a limited understanding of addition and subtraction but also in a decline in general problem-solving ability. (pp. 59-60)

It is the natural intuition of children that causes their counting strategies to bloom. Obviously, teachers need to capitalize on natural intuition when that is appropriate, and to wean children away from counting dependencies when that is appropriate. The task is not simple, but there is some guidance from the research literature.

Each child, in an individual way and at an individual pace, progresses through a sequence of counting strategies. Researchers who have studied counting (4, 10, 11) have identified six main stages:

Rote-counting. At this stage the young child has learned the number words and can recite a counting sequence, but perhaps cannot yet count objects. As Ginsburg points out (4), the term "rote" is a bit misleading, because even at this level the child develops some rules to remember that "seventeen" follows "sixteen," for example, just as "seven" follows "six."

Point-counting. Now the child can count objects while pointing to each in succession and giving each a number name. It is possible at this stage that the child obtains no meaningful information about the number of objects in the set being counted.

Cardinal-counting. For the first time, the child is able to answer quantitative questions about the set of objects being counted ("How many red blocks are there?") and answers those questions, including addition and subtraction questions, by counting *all* the objects involved, starting from 1. ("If Bill has 4 red blocks and Mary gives him 3 more red blocks, then Bill has 1, 2, 3, 4, ... 5, 6, 7 red blocks.") Ginsburg (4) points out that two prerequisites for cardinal counting are an understanding of the concept of "*one-to-one correspondence* between the counting numbers and objects (the word "one" is paired with this object, "two" with this object, etc.), and *itemization*, that is, dealing with each object in order, once and only once" (p. 110).

Counting-on. Rather than counting exclusively from 1, the child now uses the option of starting counting sequences at some point beyond 1, a breakthrough in that this skill allows work with larger numbers. When this new skill is applied to addition problems, there is usually a preliminary step of setting aside the right number of objects to be used for counting on. ("44 + 7 = ? Let's see, I'll take these seven blocks and count: [pointing] 44 ... 45, 46, 47, 48, 49, 50, 51.")

Counting-back. The child develops the ability to start at a number and count backward, opening the way to problem solving that involves subtraction. Here, as in the previous level, there is usually a preliminary step of setting aside objects. ("If Sharon has 12 apples and gives Jerry 3 apples, then she ends up with [pointing to 3 objects already counted out] 11, 10, 9 apples.")

Counting-on and counting-back directly. Although similar to the previous two levels, this level is marked by a significant step forward. Now the child does not have to set aside objects to be added or subtracted. Using fingers, other objects, or a mental tally, the child can proceed directly from the initial to the terminal numbers, as, for example, in counting 7 up from 44 to get 51, or in counting back 3 from 12 to get 9.

Furthermore, the child is now ready for open addition problems, like 13 + ? = 19.

Ginsburg (4) points out, as does the *Research Within Reach* bulletin "The Bridge from Concrete to Abstract," that at any one of these levels it is possible for a child to count without having an abstract appreciation of number. Teachers need to be alert to this possibility, because a child might fall victim to the mistaken notion that numbers are represented only by particular arrangements or structures. As an example, a child might have no clear sense that counting a row of blocks from left to right yields the same number that counting from right to left does, until she actually tries it both ways and compares answers.

When a particular child will reach these various levels of counting strategies depends very much on the child's cognitive development. Many children have the capacity to surprise those who give them the challenge, the time, and the freedom to apply their individual counting skills to relatively sophisticated problem situations (2, 5). In fact, before rushing headlong into computational procedures, primary teachers need to take a long, careful look at the skills their students *already have*. In particular, they should find out, through individual interviews, what counting strategies each child has developed ("Sharon has 9 apples after giving Jerry 3 apples? How did you get that?"). Once this diagnosis has been completed, the child's counting strategies can become the mortar for building a basic facts structure that invites both understanding and retention.

Thinking Strategies

Whether the particular focus is on counting, basic facts, or any other mathematical concept or process, the message in recent research is clear: as soon as they begin to think quantitatively, *all* children develop thinking strategies and they *use* them regularly (2, 4, 5, 10, 12). Some strategies are highly successful, some are inefficient, and some are consistently unsuccessful. The challenge for teachers is to identify these early thinking strategies, like counting, and use them in guiding children toward other thinking strategies that they will use in later mathematics.

An effective regimen for learning basic facts combines *understanding* based on concrete activities, like combining blocks or rods, with the use of *thinking strategies* to organize basic facts for retention and *drill* to ensure instant recall (7, 8). The teaching of thinking strategies appears to aid in the retention of basic facts (8, 12). Furthermore, the kinds of mathematical potential in young children Hendrickson (5) and Carpenter, Hiebert, and Moser (2) have revealed in their research provides a fertile ground for teaching thinking strategies to young children. For example, a teacher who recognizes that a particular child uses the counting-on strategy frequently and confidently might point out the additional advantages in using a strategy like *compensation*, which changes an unfamiliar fact to a more familiar fact ("9 + 6 = ? Let's see, 9 is 1 less than 10, so I can add 1 to 9 and take a 1 away from 6 [to compensate], which makes 10 + 5, which I know is 15. So, 9 + 6 = 15.").

Suggestions for helping children develop thinking strategies that are natural extensions of their self-developed counting strategies are in Rathmell (7, 8). Other suggestions that relate to the role of counting strategies in a child's mathematical growth are given below:

- Even at the preschool level, children have counting skills which teachers need to be aware of. One study of children entering kindergarten found that 75 percent could correctly rote count to ten or beyond and that 50 percent could rote count to fourteen or beyond (9). Hendrickson (5) concluded that preschool, and first-grade children could, and should, have more experience with counting objects—in particular, larger numbers of objects (more than 60 percent of the entering first graders in his study were able to count up to 18 objects).

- As you help students to develop confidence in counting-on, use *concrete activities* like dropping marbles into a container. For example, acknowledge a certain number already in the container, add 1, 2, or 3 marbles (any more is confusing to many small children), and ask how many are in the container now (1).

- If a child has difficulty keeping a mental tally when counting on from one number to another, suggest reciting the first number in a soft voice and the succeeding numbers in a louder voice (1, 6).

- Begin children on drill as soon as they show an understanding of the basic facts. Davis (3) claims that an excessive dependency on finger-counting or tally marks to figure basic facts results mainly from the failure of a child to acquire instant recall, through drill, *soon after* the facts are understood. For further information, see Davis's article and the *Research Within Reach* bulletin "Securing Mathematical Skills: Drill and Other Topics."

Conclusion

The development of counting strategies can do much more than prepare a child for computation. With teacher guidance, and supported by regular involvement in concrete activities, those strategies can mature into problem solving and mental arithmetic skills of a special quality. Indeed, since one of the primary goals of mathematics education is to train each child to think through problems to solutions, there is no better groundwork for that training than each child's own collection of thinking strategies.

References

1. Bereiter, Carl. *Arithmetic and Mathematics*. Belmont, California: Fearson Publishers, 1968.

* 2. Carpenter, Thomas P., James Hiebert, and James Moser. "The Effect of Problem Structure on First-Graders' Initial Solution Processes for Simple Addition and Subtraction Problems." Madison, Wisconsin: Wisconsin R&D Center for Individualized Schooling, 1979.

* 3. Davis, Edward J. "Suggestions for Teaching the Basic Facts of Arithmetic." In Marilyn N. Suydam and Robert E. Reys (eds.), *Developing Computational Skills*. 1978 NCTM Yearbook. Reston, Virginia: National Council of Teachers of Mathematics (NCTM), 1978.

* 4. Ginsburg, Herbert. *Children's Arithmetic: The Learning Process*. New York: Van Nostrand, 1977.

5. Hendrickson, A. Dean. "An Inventory of Mathematical Thinking Done by Incoming First-Grade Children." *Journal for Research in Mathematics Education*. 10, (January 1979): 7-23.

6. Leutzinger, Larry P., and Glenn Nelson. "Let's Do It: Counting with a Purpose." *The Arithmetic Teacher*. 27 (October 1979): 6-9.

7. Rathmell, Edward C. "How Do You Get Children to Quit Using Their Fingers to Add and Subtract?" In Mary Ellen Hynes (ed.), *Topics Related to Diagnosis in Mathematics for Classroom Teachers*. Bowling Green, Ohio: Research Council for Diagnostic and Prescriptive Mathematics, 1979.

* 8. Rathmell, Edward C. "Using Teaching Strategies to Teach the Basic Facts." In M. Suydam and R. Reys (eds.), *Developing Computational Skills*, 1978 NCTM Yearbook. Reston, Virginia: NCTM, 1978.

9. Rea, Robert E., and Robert E. Reys. "Mathematical Competencies of Entering Kindergarteners." *The Arithmetic Teacher*, 17 (January 1970): 65-74.

10. Steffe, Leslie P., and Patrick W. Thompson. "Children's Counting in Arithmetical Problem Solving." Paper presented at the Wingspread Conference on the Initial Learning of Addition and Subtraction (Racine, Wisconsin, November 26-29, 1979).

11. Thompson, Patrick W., and Alba Gonzalez Thompson. "To Count or Not to Count, Is That the Question?" Unpublished manuscript. The University of Georgia, 1978.

12. Thornton, Carol A. "Emphasizing Thinking Strategies in Basic Fact Instruction." *Journal for Research in Mathematics Education*, 9 (May 1978): 214-227.

13. Wheatley, Grayson H. "A Comparison of Two Methods of Column Addition." *Journal for Research in Mathematics Education*. 7 (May 1976): 145-154.

*The references with asterisks are readings we believe would be especially interesting to and useful for teachers.

Research Within Reach
Elementary School Mathematics

The Role of Manipulatives in Elementary School Mathematics

Question: I have a variety of manipulative aids available to me—Cuisenaire rods, base-ten blocks, counting chips, and more. For what mathematical work should my third graders be using manipulative aids?

Symbols are much more than a shorthand for recording mathematics. They provide the means for coordinating different representations of the same mathematical concept, and in so doing they can free the mind for deeper thought and quicken the pace of problem solving.

The ability to use symbols, however, is slow to develop. Piaget has charted the cognitive development of pre-adolescents, and his research indicates that even at the age of twelve, most children deal well only with symbols that are closely tied to their perceptions. For example, the symbolic representation, $1/2 + 1/3 = 5/6$, has meaning for most elementary school children only if they can relate it directly to concrete or pictorial representations.

The stages of cognitive development are dealt with in more depth in the *Research Within Reach* bulletin, "The Bridge from Concrete to Abstract." For the purposes of this bulletin, it suffices to cite the Piagetian dictum that children's learning is a process of manipulating and mentally transforming the real world. Manipulation leads to understanding and abstraction, and it is only then that a child can fully accept the symbols attached to abstractions (26).

The developmental theory of Piaget and the similar theories of Bruner and Dienes (4, 8) have intensified research interest in the role of manipulative materials (also called manipulatives) in classroom learning. Manipulatives are objects which appeal to several senses and which a student is able to touch, handle, and move. In the last three decades many studies have probed the whys, whens, hows, and whats of manipulative use. The overriding consensus is that manipulatives can help children to understand and use mathematical concepts.

In their review of the research on the use

of materials in elementary school mathematics, Suydam and Higgins noted that in almost half of the studies they considered, students who had learned with the help of manipulative materials scored significantly higher on achievement tests than students whose instruction included no use of manipulatives. In almost all of the remaining studies, the two groups were similar in their achievement; there were few instances in which the group not using materials had the higher achievement (25).

The research reveals something else about the use of manipulatives: the teacher stands at the very center of the child's experience with manipulatives, and the teacher's role is critical for the child's success. Unfortunately, it appears from a recent national survey that many teachers choose to ignore this role, and even to minimize their attention to manipulatives—especially after the primary grades. Of the K-6 teachers surveyed, nearly half reported that their students use manipulatives less than once a week, or not at all (11). There is compelling evidence from research for incorporating regular work with manipulatives into every classroom and with every child—from the elementary level on into the secondary level.

The Teacher's Role

All mathematics teaching—in measurement, numerical concepts, even in computation—should have a noticeable problem-solving flavor. That is a consistent theme throughout this series of *Research Within Reach* bulletins. Manipulatives provide an excellent means for making problem solving a pervasive influence in the classroom. From primary grades ("How many pennies do you think it will take to cover this piece of paper?") through the intermediate grades ("Can you find a way to figure out ¾ ÷ ½ by using the egg cartons?"), manipulatives invite an unlimited number of questions and challenges for any mathematical learning situation.

In fact, even before they enter first grade many children can use manipulative materials to represent the objects in problems that are stated orally. That conclusion, first noted in the 1920s, emerged once again, but with more detail, from two recent studies, one conducted by Hendrickson (12), the other by Carpenter, Hiebert, and Moser (5). Hendrickson concluded that the incoming first graders he observed could use manipulatives to "form groups, compare them, join them, separate them, and decompose them, as needed, to solve problems involving all four arithmetic operations." The four operations referred to are addition, subtraction, multiplication, and division. An example of such a problem is, "Put fourteen of your blocks in front of you. If you give me six blocks, how many blocks will you have left?"

The children in both studies were able to engage in this kind of problem solving with minimal guidance and little, if any, prior formal training. An important implication of the studies' findings is that first graders should have more opportunities to use manipulatives to represent objects in real-world problems, *before* formal training in the four arithmetic operations (12). Research has not yet delineated the process (or processes) by which children spontaneously develop an understanding of addition, subtraction, multiplication, and division, but there is no doubt that most children can develop such understanding.

In this learning process, primary grade teachers should act as guides, making a variety of manipulatives available and presenting real-world problems in the various comparing, joining, separating, grouping, and distributing operations that are related to adding, subtracting, multiplying, and dividing. Especially important for practice are situations that reinforce the notion of equivalence between a group and its constituent parts (6 is 5 and 1, 3 and 3, three 2's, and so on) (12). Finally, teachers should be asking frequent questions during this process—in particular, asking the children what they think they are being asked to do and checking to see what the thinking is behind their comparisons, groupings, and so on (12).

As they move toward abstract mathematics, children settle into their own rates of development and their own individual learning styles. For that reason, they should have a variety of manipulatives on hand throughout their elementary school experience and the freedom to choose from the available supply whenever they are doing mathematics (24, 26). Such variety and freedom of choice are essential. There is evidence, for example, that some children naturally favor *continuous* materials and others *discontinuous* materials (22). One child will grasp number concepts easily with continuous materials. Cuisenaire rods, for example, make demands on spatial perception—the orange rod represents ten, the yellow rod five, and so on. A second child might gain a clearer understanding of number by handling discontinuous materials like poker chips.

As a general rule, teachers need to be aware that children, especially primary level children, have a tendency to be distracted by irrelevant detail (23). Bana and Nelson tested the potential of manipulatives to distract children from productive thinking and found that such distractors as irrelevant colors and irrelevant numbers written on the manipulatives can indeed lead some children to false conclusions (2). However, if teachers question children regularly on sense impressions drawn from manipulatives, they can stay aware of possible misinterpretations.

Once a child has understood a particular concept using one kind of manipulative, should her teacher make sure she sees the same concept represented in other ways? For example, would there be an additional benefit for a child to see $3 \times 4 = 12$ represented with Cuisenaire rods after the representation with chips has been understood?

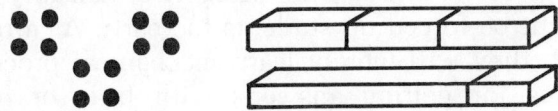

This notion of *multiembodiment* of the same concept has drawn much of its life from the writing of Dienes, who recommended multiembodiments as a smoother and more effective route to abstraction, arguing that symbols take on meaning when a child is ready to think more in terms of classes of events than single events (8). It is not clear from the research on multiembodiments whether they have any effect on immediate learning (21). In cases where there have been benefits over the long term, the key factor seems to have been the teacher's willingness to teach the children to *generalize* from one representation to another. Provided such guidance takes place, multiembodiments are recommended (25, 27).

Again and again, the role of the teacher appears as pivotal in the use of manipulatives. Even in the few studies where children who learned using a symbolic approach performed better than children who learned using manipulatives, there were signs that the children learned because of how well their teachers connected old learning with new (25). In Fennema's study, where one group of second graders learned multiplication facts using Cuisenaire rods and another group learned the same facts using a symbolic approach, the symbolic group performed better when tested on how well they could transfer their learning to unknown multiplication facts. However, both groups performed well, quite probably because the teachers of both groups taught for meaning and took pains to establish a solid background for the new concepts (9, 10).

Kinds of Manipulatives

Generally, what makes learning materials manipulative is that they appeal to several senses and are used by children for physical involvement in an active learning situation. They may be teacher-made or commercially produced; they may be environmental in nature (money, for example) or derived from mathematical structure (an abacus, for example, can provide tactile and visual reinforcement of place value). Although there is no research evidence yet to tell how many stimuli should be presented to students to maximize learning, manipulative teaching aids should include as many visual, auditory, and tactile clues as possible. The clues should be precise and fairly obvious to the children because, as we noted above, small children are easily deceived by distractions. A list of guidelines for the selection of manipulatives can be found in the article by Reys (20).

One of the more popular commercially made manipulatives has been the Cuisenaire rods.

These rods, in ten different lengths and various colors, can be used to model several mathematical concepts—number, fractions, and the four arithmetic operations are several of the more common uses. A small body of research has been directed at determining the effectiveness of Cuisenaire rods through the third-grade level. No research results are available about their effectiveness beyond the third-grade level. The available results indicate that Cuisenaire rods are more effective than traditional teaching approaches at the first-grade level, with their effectiveness diminishing in the second and third grades (10, 25). In any case, it appears that the effectiveness of Cuisenaire rods in improving a child's achievement depends on how long the child has been using the rods (10, 27). The more unfamiliar a child is with the rods, the less likely that achievement tests will favor the Cuisenaire approach over traditional approaches. In breadth of learning, it appears that Cuisenaire programs are superior to traditional programs—there is evidence that children in Cuisenaire programs learn more mathematical concepts than children taught traditionally (26).

As children move beyond the primary years, they become more adept at using symbolic mathematics. Does that mean there is no place for manipulatives in the intermediate grades? No, according to Suydam and Higgins, who summarized their survey of the research literature by writing that "the studies do not support the notion that activity lessons with manipulative materials are important at early elementary-school levels, but not at upper elementary-school levels" (25, p. 38).

Manipulatives have a place in the intermediate grades, both in developing new concepts and skills and in providing remedial help. If there is any risk related to the use of manipulatives in these grades, it derives from their being ignored or abandoned too quickly. In his book on computational error patterns, Ashlock (1) has expressed his opinion that one of the major reasons children learn patterns of error is that their teachers have introduced them to paper-and-pencil procedures while they still need to work on problems with concrete aids.

More specifically, Ashlock points to multi-digit numeration as the pivotal concept that students must learn with the help of manipulatives before they can succeed in learning computational algorithms. (A student would be well on the way to understanding multi-digit numeration, for example, if he realized that in the number 362 there are 3 hundreds, 6 tens, and 2 ones.) Ashlock then provides a series of categories of manipulatives that he thinks students should ascend through, at individual rates, to understand the concept:

Category 1: Objects which are placed in sets of ten. This may involve stacking chips, bundling sticks, or the like.

Category 2: Objects which are traded for a larger object which looks like a set of single objects glued together. Such materials include sticks used in a way where there is no bundling but only trading for already made bundles; it also includes base-ten blocks and the like.

Category 3: Objects which are traded for a single object the same size and shape as a unit object but are distinguished from the unit object by color or by both color and place. Many chip-trading activities fit this category, as does a computing abacus with a different color for each place.

Category 4: Objects traded for a single object identical to the unit object but distinguished from the unit object only by the placement of the object. Such instructional aids include many place-value charts and devices as well as a computing abacus which has discs of only one color. (1, p. 141)

Ashlock maintains that without a comfortable progression through the categories, Category 4 might be forced on students too early. As a result, they will merely learn mechanical procedures for getting answers, with little or no understanding to support them.

Manipulatives can also be valuable aids in remedial instruction. Ashlock provides a list of selected commercial materials that are appropriate for remedial instruction (1), and there are suggestions in several sources for the use of materials in remedial situations (13, 15).

Hynes, in particular, provides suggestions for corrective instruction with manipulatives in whole-number addition, subtraction, multiplication, and division. For example, he suggests that centimeter rods might prove more effective than an abacus in modelling subtraction, because the model for ten is larger than the model for one, whereas on the abacus spatial position and, perhaps, color are all that allow students to distinguish one place value from another. His model for division involves the use of graph paper by the students and a manipulative version of the so-called subtractive algorithm for division (13):

$$4\overline{)88}$$

This problem is read by the students as, "How many sets of 4 are in a set of 88?" By asking if there are 10 sets of 4 in 88, 100 sets of 4 in 88, the teacher can get the student to narrow the answer's range to somewhere between 10 and 100. Taking a section of graph paper with 88 squares on it, the student can remove 10 sets of 4.

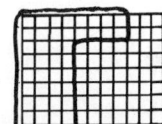

Can another set of 10 be removed? Yes.

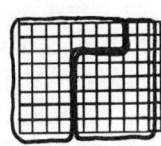 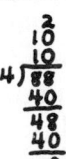

How many sets of 4 are in 8? 2 sets of 4. So the answer is 22.

If two-dimensional aids such as graph paper can benefit students, what about pictures? Are they as effective as concrete manipulatives in inducing understanding? Suydam and Higgins concluded from their survey of the research literature that "pictures are rarely superior to concrete experiences" (25, p. 24), although they do show up as superior to purely symbolic treatments in most studies that compared them (25).

Manipulatives and Selected Topics

We have already discussed the use of manipulatives in teaching concepts such as addition, subtraction, multiplication, division, multi-digit numeration, and problem solving. In this section are listed some suggestions for using manipulatives for other topics.

Counting. As soon as children start to think mathematically, they begin to develop thinking strategies. In learning to count, for example, they proceed at individual rates through a series of counting strategies, each a bit more sophisticated than its predecessor. Those strategies are described at length in the *Research Within Reach* bulletin, "Counting Strategies." In the present context, it is important to point out that manipulatives play a vital role in keeping children on course as they develop counting skills. In conjunction with teacher questions and observation, they provide the safest insurance available against the development of erroneous counting strategies.

Leutzinger and Nelson suggest a number of ways that manipulatives like counting sticks or chips can be used to reinforce the skills of counting on or counting back from one number to another (16). In the study cited previously, Hendrickson decided that many preschoolers and incoming first graders can count higher than is usually expected, and he recommended that they have more opportunities to count larger numbers of objects (12).

Part-whole relationships. This concept was mentioned in a previous section, but it deserves additional attention. Hendrickson (12) concluded that first-grade children should have more experience with groups of objects that are separated into two, three, or more parts, with emphasis on the difference between the relationship of part to part and that of part(s) to whole (8 is 2 fours, 5 and 3, and so on). Understanding part-whole relationships is an important prerequisite for learning addition, subtraction, multiplication, and division.

Money. Hendrickson also concluded from his study that incoming first graders are not as familiar with money as many textbooks and teachers assume they are. As a result, more hands-on experience with money should be available to children in the first grade (12).

Estimation. One of the cornerstones of mathematics is estimation—children should begin to develop estimating skills as soon as they begin school mathematics. They require structure and guidance in developing the skills, as well as an approach that can accommodate their need to use all their senses in their learning. With manipulatives, children can begin to learn how to estimate number ("How many marbles do you think are in this jar?") and measurement ("About how many pieces of paper will it take to cover the blackboard?"). For many more suggestions on estimation, see the article by Bright (3) and the *Research Within Reach* bulletin, "Estimation and Mental Arithmetic."

Measurement and geometry. In the recent National Assessment of Educational Progress (NAEP) in mathematics there was evidence that students do not think through measurement problems, that "they search for ways to apply the mechanical approaches that they have been taught, and if anything out of ordinary appears in the problem they are at a loss" (17, p. 62). Enslavement to mechanistic approaches begins early, and it begins as soon as a child is limited to looking at measurement problems from a distance, with no active involvement in the measuring process.

The range of manipulatives for measurement is vast, from the simple (pieces of string, coins) to the sophisticated (calipers, scales), and there is ample opportunity for teachers to develop their own devices for classroom measurement. Suggestions for the selection and design of materials can be found in the articles by Jackson and Prigge (14) and Robinson et al. (22), and in the *Research Within Reach* bulletin, "Measurement in Elementary School Mathematics."

The research into measurement is clear and emphatic in pointing to the value of *regular* hands-on measurement by children from the earliest grades. And, as noted before, if a variety of materials is available, individual styles can be better accommodated. In an experiment in teaching geometry concepts, like square, rectangle, and angle, to third graders, Prigge compared the performance of three groups—one which worked with no manipulative aids, another which worked with two-dimensional manipulatives (such as paper-folded squares and rectangles), and a third group which worked with three-dimensional manipulatives (such as cubes, tetrahedrons, and clay-formed solids). The results of one post-test and two retention tests favored the three-dimensional approach. In particular, among students who had shown up as low-ability on the Iowa Test of Basic Skills, the three-dimensional approach proved to be clearly superior to the other two approaches on tests that measured the learning and retention of geometric principles and the ability to transfer those principles to problem-solving situations (19).

The recent NAEP results add further evidence of the desirability of more hands-on learning with three-dimensional measurement manipulatives. The assessment tested students on volume, showing them a picture of a rectangular solid cut into cubes, like the one illustrated below, and asking them to find the number of cubes contained in the solid (17).

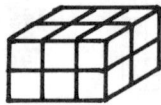

In order to solve the exercise, students had to recognize that certain cubes were not pictured in the illustration and that they must count or compute accordingly. Less than 10 percent of the nine year olds tested, less than 25 percent of the thirteen year olds, and less than 50 percent of the seventeen year olds answered the question correctly (17). The implications for more manipulative activity—in elementary school and *beyond*—are clear: teachers must provide concrete experiences with real objects, like cubes, *before* they resort to textbook pictures as the main vehicle of explanation.

The effectiveness of manipulatives in measurement instruction depends on the alertness of teachers and on their willingness to guide children past the rough spots in manipulative use. Carpenter and Lewis conducted a study with first and second graders in which they tested the children's ability to make the correct measurement conclusions when different-sized units are used to measure the same quantity (for instance, 2 quarts of liquid measured by 4 pourings with a

1-pint container and also by 8 pourings with a ½-pint container). When asked to predict how many times the smaller measuring container would have to be filled to measure the original quantity of liquid, many children were able to predict correctly that it would take more applications of the smaller measuring unit than of the larger unit. However, when both units were used to measure out equal quantities of liquid, many of those same children compared the resulting quantities and concluded that they were unequal (6).

The researchers concluded that manipulatives with different units of measure might interfere with a young child's intuitive understanding of the relationship between the size of a unit of measure and the number of times it must be used to measure a given quantity. Between the ages of four and eight, children are especially susceptible to such logical inconsistencies, and teachers need to be alert for them.

Conclusion

Symbols become appropriate after their associated concepts have been abstracted. The process of mathematical abstraction begins for children in their interactions with their environment, then with concrete manipulatives that guide them toward mathematical concepts. Manipulatives not only show the way to conceptual understanding, but they provide experiences in which children can transfer their understanding smoothly from one concept to another. By their appeal to a variety of senses, they also motivate children.

Intriguing as they are at first, and though they have a role in learning and instruction past the elementary school years, manipulatives are by their nature cumbersome for problem solving, and it is only natural that children develop more symbolic approaches to problem solving as their cognitive development advances. Research into child development shows that the development from concrete to abstract has a basic inevitability about it. But the research into manipulatives and mathematics instruction shows that the *quality* of learning is not inevitable and that children cannot learn mathematics well with manipulatives alone. For an experience with mathematics that is firmly planted and easy to build on, children need the guidance of aware, informed, and caring teachers.

References

* 1. Ashlock, Robert B. *Error Patterns in Computation: A Semi-Programmed Approach.* Columbus, Ohio: Charles E. Merrill Publishing Co., 1976.

2. Bana, Jack, and Doyal Nelson. "Distractors in Nonverbal Mathematical Problems." *Journal for Research in Mathematics Education*, 3 (November 1972): 233-238.

* 3. Bright, George W. "Estimation as Part of Learning to Measure." In Doyal Nelson and Robert E. Reys (eds.), *Measurement in School Mathematics*, 1976 NCTM Yearbook. Reston, Virginia: National Council of Teachers of Mathematics (NCTM), 1976.

4. Bruner, Jerome. *Toward a Theory of Instruction.* New York: W.W. Norton, 1966.

5. Carpenter, Thomas P., James Hiebert, and James M. Moser. "The Effect of Problem Structure on First Graders' Initial Solution Processes for Simple Addition and Subtraction Problems." Madison, Wisconsin: Wisconsin R&D Center for Individualized Schooling, 1979.

6. Carpenter, Thomas P., and Ruth Lewis. "The Development of the Concept of a Standard Unit of Measure in Young Children." *Journal for Research in Mathematics Education*, 7 (January 1976): 53-58.

* 7. Cathcart, W. George (ed.). *The Mathematics Laboratory: Readings from the Arithmetic Teacher.* Reston, Virginia: NCTM, 1977.

8. Dienes, Zoltan P. "Some Basic Processes Involved in Mathematics Learning." In Robert B. Ashlock and Wayne L. Herman, Jr. (eds.), *Current Research in Elementary School Mathematics.* New York: The MacMillan Company, 1970.

9. Fennema, Elizabeth H. "The Relative Effectiveness of a Symbolic and a Concrete Model in Learning a Selected Mathematical Principle." *Journal for Research in Mathematics Education*, 3 (November 1972): 233-238.

10. Fennema, Elizabeth H. "Models and Mathematics." *Arithmetic Teacher*, 19 (December 1972): 635-640.

References (continued)

11. Fey, James T. "Mathematics Teaching Today: Perspectives from Three National Surveys." *Arithmetic Teacher*, 27 (October 1979): 10-14.

12. Hendrickson, A. Dean. "An Inventory of Mathematical Thinking Done by Incoming First-Grade Children." *Journal for Research in Mathematics Education*, 10 (January 1979): 7-23.

* 13. Hynes, Michael C. "Using Manipulative Aids to Model Algorithms in Remedial Situations." In Mary Ellen Hynes (ed.), *Topics Related to Diagnosis in Mathematics for Classroom Teachers*. Bowling Green, Ohio: Research Council for Diagnostic and Prescriptive Mathematics, 1979.

* 14. Jackson, Robert L., and Glenn R. Prigge. "Manipulative Devices with Associated Activities for Teaching Measurement to Elementary School Children." In D. Nelson and R.E. Reys (eds.), *Measurement in School Mathematics*, 1976 NCTM Yearbook. Reston, Virginia: NCTM, 1976.

15. Junge, Charlotte W. "Adjustment of Instruction (Elementary School)." In *The Slower Learner in Mathematics*, 35th NCTM Yearbook. Reston, Virginia: NCTM, 1972.

16. Leutzinger, Larry P., and Glenn Nelson. "Let's Do It: Counting With a Purpose." *Arithmetic Teacher*, 27 (October 1979): 10-16.

17. National Assessment of Educational Progress. *Mathematical Knowledge and Skills: Selected Results from the Second Assessment of Mathematics*, Report No. 09-MA-02. Denver, Colorado: Education Commission of the States, 1979.

18. Nelson, Doyal, and Robert E. Reys (eds.). *Measurement in School Mathematics*, 1976 NCTM Yearbook. Reston, Virginia: NCTM, 1976.

19. Prigge, Glenn R. "The Differential Effects of the Use of Manipulative Aids on the Learning of Geometric Concepts by Elementary School Children." *Journal for Research in Mathematics Education*, 9 (November 1978): 361-367.

* 20. Reys, Robert E. "Considerations for Teachers Using Manipulative Materials." *Arithmetic Teacher*, 18 (December 1971): 551-558.

21. Reys, Robert E. "Mathematics, Multiple Embodiment, and Elementary Teachers." *Arithmetic Teacher*, 19 (October 1972): 489-493.

* 22. Robinson, G. Edith, et al. "Measurement." In Joseph N. Payne (ed.), *Mathematical Learning in Early Childhood*, 37th NCTM Yearbook. Reston, Virginia: NCTM, 1975.

23. Sharma, Mahesh. "The Problem of the 'Missing Addend.'" In *Math Notebook*. Framingham, Massachusetts: The Center for Teaching/Learning of Mathematics, 1979.

24. Stevenson, Harold W. "Learning and Cognition." In J.N. Payne (ed.), *Mathematics Learning in Early Childhood*. 37th NCTM Yearbook. Reston, Virginia: NCTM, 1975.

* 25. Suydam, Marilyn N., and Jon L. Higgins. *Activity-Based Learning in Elementary School Mathematics: Recommendations from Research*. Reston, Virginia: NCTM, 1977.

26. Suydam, Marilyn N., and J. Fred Weaver. "Research on Mathematics Learning." In J.N. Payne (ed.), *Mathematics Learning in Early Childhood*. 37th NCTM Yearbook. Reston, Virginia: NCTM, 1975.

27. Suydam, Marilyn N., and J. Fred Weaver. *Using Research: A Key to Elementary School Mathematics*. Reston, Virginia: NCTM, 1975.

*The references with asterisks are readings we believe would be especially interesting to and useful for teachers.

Research Within Reach
Elementary School Mathematics

Measurement in Elementary School Mathematics

Question: How should measurement be taught? How early in the curriculum should metric units be introduced?

Whether things are measured in yards or meters is of little importance in the education of a young child. In either case, the concepts and skills of measurement are the same. For adults, and for teachers in particular, the prospect of a change in measurement systems can be disquieting. At the same time, it provides a chance for teachers to approach measurement instruction with renewed interest and a fresh perspective.

There are specific *skills* and *concepts* children should learn in measurement, whether the language is metric or traditional. As they develop the appropriate skills and conceptual understanding, they must pass through a sequence of measurement processes, determined in part by the pace of cognitive development. This bulletin will highlight those skills, concepts, and processes and will also address the issue of metric education.

Overview of Measurement Concepts and Skills

Our society measures a variety of attributes in a variety of ways. As a framework for later recommendations from research and practice, here is a quick overview of the concepts and skills involved.

Things we choose to measure have several attributes (17):

area	length	speed
capacity	pressure	temperature
density	volume	viscosity
distance	weight	

Slightly removed from this list, because it is not an attribute of things as such, is the equally important concept of *time*.

Serving as a connective concept for all of these is the concept of *measurement unit* (inch, kilogram, etc.). An analysis of the 1972 National Assessment of Educational Progress suggests that a weakly developed understanding of units lies at the root of the poor showing by many students in the Assessment's area, volume, and temperature exercises (4). Although the concept of units is not one most children learn quickly, it is vital to the learning of measurement, and teachers must carefully nurture their students' appreciation of units.

Various measuring *skills* help children to understand and work with measurement concepts, and all of these skills deserve regular and careful attention in the classroom. For example, a two-dimensional figure suggests several approaches to area measurement:

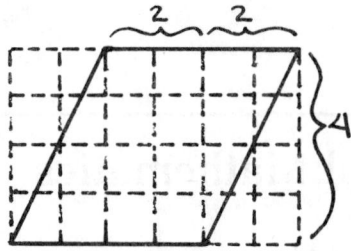

Estimation—"The figure is more than ½ the size of the 4×6 rectangle, which has an area of 24 square units."

Covering by units and counting—"I am going to see if I can count how many little unit squares cover the figure."

Partitioning and recombining—"If I remove the triangle on the left side of the parallelogram and put it on the right side, I'll get a rectangle with the same area as the parallelogram; then I'll be able to figure it out, because I know how to compute the area of a rectangle."

Computation—"I can use the formula for the area of a parallelogram, area = base × vertical height."

Other measurement situations call for different measuring skills. For example:

Counting—"This bag must weigh twice as much as that one because it has twice as many marbles in it."

Indirect measure—"Which is longer, that metal rod or the distance around this cylinder? I can find out by using a piece of string to measure off the cylinder, then hold the string next to the rod and compare."

Using instruments—"This measuring tape tells me that the belt is 60 centimeters long."

The Development of Measurement Processes

The above concepts and skills can translate quite readily into instructional goals. However, the dictates of cognitive development make it impossible for children to reach those goals either quickly or in unison. In fact, Piaget says that small children may believe that the length of a measuring stick changes as it is shifted (17). In light of that belief, it would be unwise to rush children into expecting *accuracy* in their measurements.

Research has also shown *transitivity* to be a prerequisite for understanding units of measurement (18). An example of transitivity is, "if A is smaller than B, and B is smaller than C, then A is smaller than C."

Transitive reasoning is evident in some children at ages five and six (18), but approximately 30 percent of children still do not have a firm grasp on transitivity at age eight (1). Clearly, these youngsters cannot fully appreciate the concept of a measurement unit until they have such a grasp.

On the other hand, over 50 percent of entering kindergarteners can differentiate between the characteristics of size and weight (16). A question arises about the kinds of measurement experiences that can build on these early foundations, while at the same time not souring the gradual and clearly individual growth of children?

It might help to view the process of learning measurement as having four parts, with the child proceeding from *perception* to *comparison* to measurement with *nonstandard measures*, and finally, to the use of *standard measures* (11).

Perception. "Perception is readiness; it is vocabulary building; and it is a foundation for understanding" (11). Teachers of young children should regularly appeal to children's perception of all the attributes listed earlier in this bulletin. Perception, not *number*, should be the focus during this period—for example, with the children watching, weigh objects that are familiar to them, ask for ideas on how the group might find out how tall each child is, and so on. (See references 11, 15, and 17 for many more suggestions.)

Comparison. Once children are proficient in perceiving measurement attributes, they should practice comparing those attributes ("Which will require more paper to cover, the bulletin board or the door? How could you find out?" "Who is taller, Margaret or Barbara? How could you find out?"). As much as possible, pose measurement questions in *problem* settings, where the children answer questions by deciding if they have "enough" of something, or if they have "too much" or "too little," or if something is "too long" or "too short." See the article by Robinson (17) for further suggestions.

Nonstandard measurement. Before children begin to work with inches, centimeters, and so on, teachers have the opportunity to help them relate the measurement process to their immediate environment by using convenient nonstandard measures. ("It takes 10 pieces of construction paper to cover the bulletin board." "The length across the floor is 30 lengths of Bob's feet.") Several people who have studied the teaching and learning of measurement regard experiences in nonstandard measurement as a good problem solving investment—children can develop their own resourcefulness in measuring and avoid a possible later dependence on measuring instruments (4, 17).

A note of warning: this is a period in their development when many children are prone to misunderstandings about the measurement process. A look at some common misconceptions appears later in this bulletin, but in the present context teachers should be aware of the confusions that might arise for children if, for example, they are told that it takes 30 lengths of Bob's feet to measure the width of the classroom, and 40 lengths of Gloria's feet to measure the same distance. To an adult, that just means that Gloria's feet are smaller; to a young child, the implications might not be clear at all.

Standard measurement. The move from nonstandard to standard measurement can be gradual or nearly simultaneous. What is important is that the children are doing measurement on a regular basis. One transition route is class discussion of the recording and communication problems that arise if, to use the example above, it takes 40 Gloria-footlengths to cover the same distance as 30 Bob-footlengths.

The introduction and pacing of these four processes must always depend on children's levels of development, but one extensive school study gives evidence that perception and comparison should be handled in the lower grades (K-2), that children can profit most from nonstandard measures at or before the third-grade level, and that teachers should develop student appreciation of standard measures at or above the third-grade level (11).

Teaching Measurement Concepts and Skills

The two National Assessments of Educational Progress in mathematics (NAEP, 1974 and 1978) revealed measurement as an area where children's skills are weak. Other studies have isolated measurement misconceptions that are common among elementary school children. A close look at this information provides suggestions for sustaining and reinforcing measurement concepts and skills (4, 12).

In both NAEP assessments there were glaring deficiencies that showed up in the length, area, and volume problems. For example, only 7 percent of nine year olds in 1974 were able to find the amount of fencing needed to enclose a rectangular garden nine feet long and five feet wide (4). In the most recent assessment, however,

40 percent of nine year olds responded correctly when asked (12)

What is the distance all the way around this rectangle?

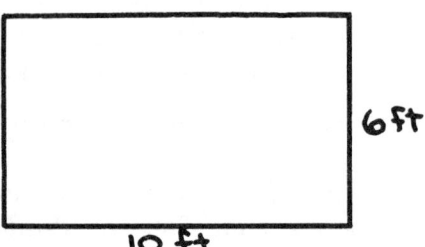

Why the increase from 7 to 40 percent? It could be that children need more practice in *thinking through* measurement problems, like the first one, that lack direct instructions. It is also quite possible that many children grasp the concept of perimeter but are weak in vocabulary ("enclose," "rectangular").

In both assessments most nine year olds could measure accurately with a ruler only if the length they were measuring was shorter than the ruler and was represented by a whole number on the ruler (4, 12).

Area and volume problems were especially difficult, with fewer than 30 percent of nine year olds able to compute successfully the area of a rectangle in which unit squares had been drawn (4). As noted above, research studies have implied that a majority of nine year olds should be developmentally ready to understand covering by units. Why should such a discrepancy show up on the tests? One probable reason: Children do not get enough practice in *meaningful* measurement computation, practice that connects the computation to the covering by units.

One group of researchers (10) set out to determine common misconceptions about area among children in grades three through six. As the researchers pointed out, the implications go beyond measurement as an isolated topic—a faulty understanding of area can interfere with a child's appreciation of the visual representations of other mathematical concepts. For example, if a child doesn't understand the concept of area, then it makes no sense to illustrate the concept of fraction with an area diagram.

Using the length of one dimension to make area judgments. A child whose attention is isolated on the base dimensions might decide that the rectangle on the left has a larger area:

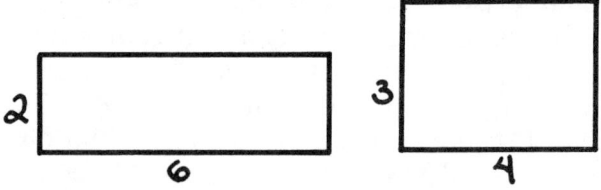

Using primitive compensation methods. A child might begin with the healthy insight that to make a smaller rectangle equal in area to a larger rectangle she has to extend the smaller rectangle. However, her insight falters at the point of deciding *how many* units to add on.

Point-counting for area. A child might rely too heavily on computing area by pointing to each unit square and counting. This habit can carry the child successfully until he attempts to solve a problem that involves fractional parts. ("Make a rectangle that is 4 units wide and has an area of 18 square units.")

Counting around the corner. A child with this misconception would say that the area of the following figure is 8:

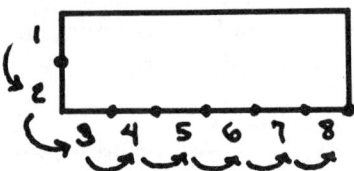

Point-counting endpoints. This is a frequently observed misconception. Here the child will count the *points*, not the units, along one dimension of a figure, giving an answer, perhaps, of 20 for the area of the figure below:

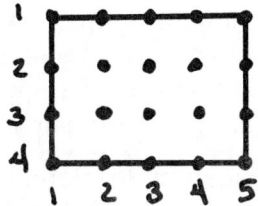

One recommendation emerging from research is that teachers delay stressing multiplication (area = length × width) as a way to compute area until their students are familiar with computing area *both* by covering/counting by units *and* by partitioning/recombining (10).

Both NAEP assessments tested students on volume problems, and the results in both assessments were striking. For example, the students were shown a picture of a rectangular solid cut into cubes, like the one illustrated below, and were asked to find the number of cubes contained in the solid (12).

In order to solve the exercise, students had to recognize that certain cubes were not pictured in the illustration and that they must count or compute accordingly. Less than 10 percent of the nine year olds, less than 25 percent of the thirteen year olds, and less than 50 percent of the seventeen year olds, on *both* assessments, answered the question correctly. The implications seem clear: in developing the concept of volume, teachers must provide concrete experiences with real objects *before* they resort to textbook pictures as the main vehicle of explanation.

Researchers have discovered a few other measurement difficulties that children have, difficulties which are connected to the concept of length and derived mainly from the limitations of cognitive development. Children in the first and second grades (and some in later grades) are easily confused by a switch in measurement units. For example, even if they see that two objects of the *same* length are about to be measured, if centimeter units are used to measure the one and decimeter units the other, the children are liable to succumb to their size-of-units/number-of-units confusion and decide that the original lengths were *not* the same (5). The implication: teachers should not rush to have students do manipulations with *different-sized units*. (Note that this research underscores the warning attached to our earlier description of nonstandard measurement. There is some value in children discussing how many lengths of Bob's feet it will take to measure a classroom, but they may get confused if they are faced with the statement that it takes 40 of Gloria's foot-lengths to cover the same distance.)

Another study discovered that most children do not move beyond this difficulty until after they are nine years old (1). Here there is an implication for the teaching of fractions by use of a number line. A nine-year-old or younger child might be hard pressed to explain why 5/7 appears to the left of 4/5; there could be some confusion about the number of units versus the number of times they are counted—four of one unit cover more of the number line than five of the other:

Besides watching for misconceptions and observing the limits set by cognitive development, what else can teachers do to enrich their teaching of measurement? There are several suggestions, the first of which highlights a teaching aid with a tremendous potential to inspire learning—if it is practiced *regularly*. That teaching aid is estimation.

Estimation and Measurement

Bright (2) gives a comprehensive portrayal of the role of estimation in the teaching and learning of measurement. From his analysis, there arise eight basic kinds of estimation. When that eight is multiplied by the number of attributes for measurement (length, area, weight, and so on), a wealth of ideas for classroom discussion emerge.

According to the analysis, four of the basic kinds of estimation arise when a measure for a given object is to be estimated ("Draw a diagonal on this piece of paper. Estimate its length to the nearest centimeter."). The other four arise when a measure is given and the goal is to name an object for which that measure is a reasonable estimate ("Name something that is 70 centimeters long.").

We recommend that you read Bright's article in its entirety. With the framework that Bright provides, a teacher can establish and maintain the balance of measurement activities that is so crucial in concept development. We also recommend the *Research Within Reach* bulletin "Estimation and Mental Arithmetic."

Other Recommendations

Besides recommending regular instruction and practice in estimating measurements, the researchers who reviewed the NAEP results made the following recommendations (4):

- Give particular attention to the development of the concept of unit.

- Provide a wide variety of measurement experiences at all grade levels.

- Give instruction and regular practice in measuring lengths that are longer than the measuring instrument; making measurements that involve fractions other than one-half; and making indirect measurements of distance (see references 11 and 17 for suggestions).

- Provide problems involving intervals of time rather than merely giving practice in clock-reading skills. ("Bill went to work at 8:45 and returned 10 hours later. At what time did he return?")

- Use everyday situations that emphasize the notion that measurement is one of the principal ways mathematics is applied in the real world.

- Use regular questions to prepare your students for *thinking through* measurement problems. ("If you know that one side of a square is 5 centimeters, how can you find the area?")

In the area of curriculum development, one program deserves special note because of its regular attention to measurement. Designed for grades K-6, the Developing Mathematical Processes (DMP) program uses linear measurement as an early and systematic vehicle for developing skills in comparing, classifying, and ordering. The program continues to use measurement, especially hands-on measurement of length, weight, angles, and so on, as a way to develop more sophisticated mathematical concepts and skills. Interested teachers can obtain more information from the distributors of the program (6).

Metric Education

There are numerous sources of guidelines for teaching metric measurement (for example, references 8, 9, and 13). In general, the recommendations are consistent with the picture research has sketched of the child's growth in understanding measurement. Work in kindergarten through grade two should highlight perception and comparison. At the third-grade level, or a bit earlier, metric units can enter as a supplement to nonstandard measures. Linear units (meter, centimeter, millimeter) should come first, but teachers should choose and concentrate for a while on *one unit*, minimizing the use and conversion of multiple units. Educators differ on what that unit should be (8), but the message is clear: once you choose, be consistent.

Regular work with *estimation* and *actual measuring* reinforces and sustains what the child learns, and increases motivation ("Which comes closest to the width of this desk? 10 cm, 50 cm, 100 cm? Why? Check and see."). The metric system and our traditional system should be treated separately in the curriculum (13), and teachers should give *very little* attention to conversion between the two systems. If attention *is* given to conversion, it should take the form mostly of approximations (for example, "one meter is a little more than a yard") (13).

Conclusion

Introducing the metric system into our schools has not, and will not, change the way children learn measurement. Their needs are tied to individual cognitive development and thus transcend particular systems of measurement. The process of matching instruction with changing developmental needs has become clearer through research, but the challenge of clarifying the process of learning measurement continues, both for researchers and for teachers.

References

1. Bailey, Terry G. "Linear Measurement in the Elementary School." *Arithmetic Teacher*, 21 (October 1974): 520-525.
* 2. Bright, George W. "Estimation as Part of Learning to Measure." In Doyal Nelson and Robert E. Reys (eds.), *Measurement in School Mathematics*, 1976 NCTM Yearbook. Reston, Virginia: National Council of Teachers of Mathematics (NCTM), 1976.
3. Carpenter, Thomas P. "Analysis and Synthesis of Existing Research on Measurement." In Richard A. Lesh (ed.), *Number and Measurement*. Columbus, Ohio: ERIC Clearinghouse for Science, Mathematics, and Environmental Education (ERIC/SMEAC), 1975.
4. Carpenter, Thomas P., et al. *Results from the First Mathematics Assessment of the National Assessment of Educational Progress*. Reston, Virginia: NCTM, 1978.
5. Carpenter, Thomas P., and Ruth Lewis. "The Development of the Concept of a Standard Unit of Measure in Young Children." *Journal for Research in Mathematics Education*, 7 (January 1976): 53-58.
6. *Developing Mathematical Processes*. Chicago, Illinois: Rand McNally and Company. (P.O. Box 7600, Chicago, IL 60680) This program was developed during the early 1970s by the Wisconsin R&D Center for Individualized Schooling.
* 7. Downes, John P., et al. *76 Questions: A Synthesis of the Research on the Teaching and Learning of Mathematics*. Atlanta, Georgia: Research and Development Utilization Project, Georgia Department of Education, 1977. [ED 162 896]
8. Goldbecker, Sheralyn S. *Metric Education*. Washington D.C.: National Education Association, 1976.
9. Higgins, Jon L. (Ed.). *A Metric Handbook for Teachers*. Reston, Virginia: NCTM, 1974.
10. Hirstein, James J., et al. "Student Misconceptions about Area Measure." *Arithmetic Teacher*, 25 (March 1978): 10-17.
* 11. Inskeep, James E., Jr. "Teaching Measurement to Elementary School Children." In D. Nelson and R.E. Reys (eds.), *Measurement in School Mathematics*, 1976 NCTM Yearbook. Reston, Virginia: NCTM, 1976.
12. National Assessment of Educational Progress. *Mathematical Knowledge and Skills: Selected Results from the Second Assessment of Mathematics*. Report No. 09-MA-02. Denver, Colorado: Education Commission of the States, 1979.
13. NCTM Metric Implementation Committee. "Metric: Not If, But How." *Arithmetic Teacher*, 21 (May 1974): 366-369.
* 14. Nelson, Doyal, and Robert E. Reys (Eds.). *Measurement in School Mathematics*. 1976 NCTM Yearbook. Reston, Virginia: NCTM, 1976.
15. Pagni, David L. "Applications in School Mathematics: Human Variability." In Sidney Sharron and Robert E. Reys (eds.), *Applications in School Mathematics*, 1979 NCTM Yearbook. Reston, Virginia: NCTM, 1979.
16. Rea, Robert E., and Robert E. Reys. "Competencies of Entering Kindergarteners in Geometry, Number, Money, and Measurement." *School Science and Mathematics*, 71 (May 1971): 389-402.
* 17. Robinson, G. Edith, et al. "Measurement." In Joseph N. Payne (ed.), *Mathematics Learning in Early Childhood*, 37th NCTM Yearbook. Reston, Virginia: NCTM, 1975.

References (continued)

18. Steffe, Leslie P., and James J. Hirstein. "Children's Thinking in Measurement Situations." In D. Nelson and R.E. Reys (eds.), *Measurement in School Mathematics*, 1976 NCTM Yearbook. Reston, Virginia: NCTM, 1976.

*The references with asterisks are readings we believe would be especially interesting to and useful for teachers.

Mathematical Development: The Teacher's Role

Research Within Reach
Elementary School Mathematics

Diagnosis: Taking the Mathematical Pulse

Question: At the start of every school year I am faced with the same concern: What do my students know? Where are each student's weak and strong areas in mathematics?

Include "diagnosis" in a word association game and you invite immediate responses of "doctor," "illness," and the like. It is natural, then, that when you restrict the word to a mathematics classroom there will be a tendency for teachers to associate "diagnosis" with learning sickness and remediation.

There is no doubt that the teacher who is adept at diagnosis is a person with a keenly developed eye for persistent trouble spots in a child's mathematical growth, a deft touch in probing for the causes of the trouble, and, possibly, an awareness of ways to deal with deficiencies. But if the target focus is widened, if the teacher sets out in the beginning of the school year to assess each child's mathematical strengths, as well as weaknesses, then there will be additional payoffs for both teacher and child.

The research literature lends support to your giving mathematical diagnosis a double scope—strengths as well as weaknesses. Indeed, in cases where weaknesses have been isolated, researchers recommend that you involve the student in planning work in the weak areas (1). A firm foundation for this involvement lies in acknowledgment of the student's areas of mathematical strength by *both* teacher and student. For example, a child who has trouble with the steps of a multiplication algorithm may be competent in estimation. Another child who is weak in recalling basic facts may have developed impressive counting strategies to compensate. Building on skills such as these, teachers can help students deal with their deficiencies in mathematics—by making sure, for example, that the child who estimates well recognizes the value of her skills and that she has time, and takes time, to estimate before each multiplication exercise.

There are several ways in which you as teacher can take your students' mathematical pulse, and the range of options can fit any personal style or classroom design. Loosely categorized, the ways of classroom diagnosis are observation, personal interview, and paper-and-pencil testing. Observation and interview can reveal child behaviors not noticeable from paper-and-pencil tests, while testing, on the other hand, will sometimes provide more convenience and a wider range of investigation. Thus, some blending of the three formats often makes good sense.

At this point, we list research guidelines for diagnosis in elementary school mathematics, including references to articles and books you might find helpful in organizing diagnostic interviews and tests.

Developmental and Motivational Factors

- Make sure that a child's apparent mathematical deficiency is really a deficiency. For example, Riedesel (16) found many children who did not know basic addition facts but who did know basic multiplication facts. The evidence indicated that the typical attitude was that there is not much time saved in memorizing 9 + 8 = 17, but it takes too long to figure out 9 × 8 = 72. Thus, what appeared to be an inability to memorize basic facts turned out to be no such thing.

- Remember, furthermore, that each child progresses through several stages of development before reaching an adult conceptual level in mathematics. Occasionally you may see a seven-year-old child who shows a better sense of number, for example, than an eight-year-old classmate, for the sole reason that the younger child has developed faster.

  A comprehensive source of information on the mathematical development of young children, with practical classroom suggestions, is the 37th yearbook of the National Council of Teachers of Mathematics (NCTM), *Mathematics Learning in Early Childhood* (14).

- Strengthen your diagnosis, whether from observation, interview, or testing, with the liberal use of manipulatives. The value of manipulatives extends far beyond the bounds of diagnosis, but in light of the role that development plays in diagnosis, it is important that you assess a child's understanding of a concept or operation at concrete, pictorial, and abstract levels.

  For example, suppose the following type of subtraction error appears regularly in a child's work:

  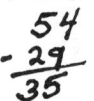

  If you are left unsure whether its root cause is carelessness, a lack of understanding of the base-ten numeration system, or even a lack of understanding of the concept of subtraction, then it would be advisable for you to ask the child to represent a similar exercise with base-ten blocks (or bundles, rods, an abacus, etc.). The problem then becomes one of removing 2 tens blocks and 9 ones blocks from the total of 5 tens blocks and 4 ones blocks. The child's manipulation will give you a much sharper focus on her thinking.

  This willingness and readiness to ask "Can you show me without using pencil and paper?" or "Can you show that with a picture?" is a powerful diagnostic ally throughout the elementary school years, and, indeed, beyond.

- Don't lose sight of the emotional side of students in your diagnosis. Denmark relates his experience with a fifth-grade child, presumed to have no understanding of division, even on the concrete level, who became a division dynamo when Denmark persisted in his diagnosis and introduced a game involving division. His assessment was that the child's needs probably lay in the area of motivation (6).

  This example illustrates a crucial component in diagnostic teaching—you must be both flexible and patient in piecing together an accurate picture of a child's thinking.

- A second, equally crucial, component is a climate of acceptance (1). Effective diagnosis requires two-way communication, and a child is unlikely to cooperate in revealing his thinking unless he perceives not only that you are willing to help him, but that you stand ready to accept his failures and incorrect responses without criticism.

Further recommendations for motivating and establishing trust in students are in our *Research Within Reach* report "Motivation in Mathematics." Examples of games with diagnostic potential can be found in *Didactics and Mathematics* (5) and Johnson (11).

Diagnostic Testing

Models for diagnostic tests in mathematics can be found in Reisman (15), Underhill (18), Johnson (11), and Herold (8). Reisman also includes guidelines for designing your own diagnostic tests. However, if you do set out to design your own diagnostic test, consider the following points:

- Take care to construct exercises that will allow you to distinguish basic facts ignorance from algorithm difficulty. For example, Inskeep (9) provides a series of exercises intended to test understanding of multidigit multiplication. For the design of the exercises he stresses the importance of control over the basic facts involved. As an example, one might avoid using 7, 8, or 9 as digits, thus bypassing the basic facts most students find most difficult and clearing the way for the teacher to judge the degree of each child's mastery of the multiplication algorithm.

Inskeep recommends giving a basic facts test before your diagnostic test, including only those facts used on the diagnostic test.

- Distinguish, in your diagnostic evaluations, between computational errors that are random and those which occur more systematically. Systematic errors are more deeply rooted than random errors, as Cox illustrated with her report that 23 percent of the children she tested who were making systematic computational errors were still making the same or very similar errors almost one year later. In her study she defined a systematic error as one which occurs "in at least three out of five problems for a specific algorithmic computation" (4).

Despite the tendency of systematic errors to stick with a child for a long time, it appears that remedial intervention can hasten their elimination. It will pay you, therefore, to make the effort to isolate them. The example below shows evidence of a systematic error:

(1) $\begin{array}{r}476\\+17\\\hline 25\end{array}$ (2) $\begin{array}{r}205\\+86\\\hline 21\end{array}$ (3) $\begin{array}{r}754\\+28\\\hline 26\end{array}$

The answers above come from adding up *all* the digits in the addends—e.g., 25 = 4 + 7 + 6 + 1 + 7.

As you work to make yourself more alert to systematic errors, you may find it helpful to read Backman's article in the NCTM 1978 yearbook, in which he provides a list of computational errors and thought patterns of children which may enrich your own diagnostic observations (2). In particular, he analyzes error patterns by attending to:

1. Errors related to conceptual learning:

$$\begin{array}{r}402\\\times\ 6\\\hline 2472\end{array}$$

The student believes 6 × 0 = 6.

2. Errors related to sequencing steps within procedures:

$$4\overline{)418}^{\,142}$$

The student reversed the role of dividend and divisor in part of the division.

3. Errors related to selecting information or procedures:

$$\begin{array}{r}53\\\times 2\\\hline 105\end{array}$$

The student used an addition fact, 2 + 3 = 5, within the multiplication procedure.

4. Errors related to recording work.

As you develop skills in recognizing systematic errors and the thinking patterns that produce them, it is important to consider *why* children learn patterns of error. There is no one cause, but a major contributing factor appears to be the premature introduction of a child to paper-and-pencil procedures while he still needs to be working with concrete aids (1).

Other Diagnostic Concerns

- Be concerned, in your diagnosis, with mathematical areas which are not primarily computational. For example, the last National Assessment of Educational Progress (NAEP) in mathematics included a test exercise in area:

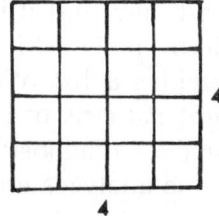

Which of the figures below has the same area as the figure above?

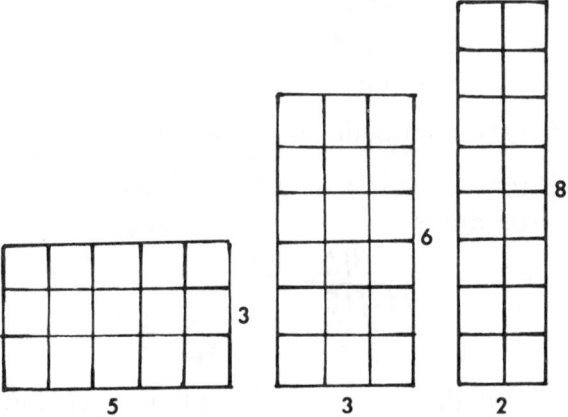

Those who reported on the NAEP results concluded:

> Thus, it appears that most 9 year olds do not intuitively recognize the consequences of partitioning regions into units of measure. This principle is one of the first that must be mastered for any subsequent work on measuring area to have any meaning. (3)

The diagnostic value of such exercises becomes clear: if you as a teacher do not take pains to see if partitioning by units makes sense to your students, then you may be left, at best, with mere rote learning if you decide to move on to procedures for calculating areas.

Diagnostic Interviews

- Use individual interviews whenever your observation or diagnostic testing leaves you confused about a particular child's mathematical thinking. In the recommended format (13), teacher and student meet together away from the rest of the students, but not necessarily out of the classroom. The student works through a set of exercises and is asked to make clear her thinking, out loud, either while she is working or after the work is complete.

Be as unobtrusive as possible when you interview, limiting your interventions to ensuring that the student does clarify her reasoning and strategies. The interview is *not* a teaching exercise, so you should ignore "Is that right?" questions from the child and also avoid nonverbal, as well as verbal, signals to the student. Adapt your pace of interviewing to the pace at which each child is comfortable responding, allow the child to use his own language, imprecise though it may be, and be ready to stop an interview as soon as the child shows anxiety, rebelliousness, or boredom.

Prerequisites

Once your diagnosis has helped you to isolate a concept that a particular child doesn't understand or a skill she hasn't mastered, and before you embark on remedial work, it is necessary to do some tracing back to assess where the understanding became faulty. There are a number of available lists of prerequisite hierarchies for elementary school mathematics; one such list can be found in Reisman's book (15).

However, it is possible for you to do some tracing back on your own. To generate your own list of prerequisites for a particular skill or concept, you need to give careful thought to the question, "What would a student need to have learned in order to understand this concept (or have this skill)?"

For example, if you have concluded that a student doesn't understand place value tasks involving hundreds, tens, and ones, then the answer to the above questions might involve 4 steps. The student would need to have learned to

1. be able to order numbers from 100-999;
2. interpret place value of a three-place numeral;
3. write numerals in the hundreds;
4. exchange tens for hundreds and hundreds for tens (e.g., 17 tens is 1 hundred, 7 tens).

When you have done this, it is possible to assess the child's understanding of each of these prerequisites, and to carry on, if necessary, the process of tracing back.

Conclusion

Diagnosis, properly done, can be a powerful teaching ally. With it, you can draw a profile of each student for areas of possible weakness, areas of consistent weakness, areas of transitional growth, and areas of actual and potential strength. Then, as you face each child's individual needs, you will be better armed to meet those needs.

References

* 1. Ashlock, R.B. *Error Patterns in Computation.* Columbus, Ohio: Charles E. Merrill, 1972.
 2. Backman, Carl A. "Analyzing Children's Work Procedures." In M. Suydam and R. Reys (eds.), *Developing Computational Skills,* 1978 NCTM Yearbook. Reston, Virginia: National Council of Teachers of Mathematics (NCTM), 1978.
 3. Carpenter, Thomas P., et al. "Notes from National Assessment: Basic Concepts of Area and Volume." *The Arithmetic Teacher,* 22, 6 (October 1975): 501-507.
 4. Cox, Linda S. "Diagnosing and Remediating Systematic Errors in Addition and Subtraction Computations." *The Arithmetic Teacher,* 22, 2 (February 1975): 151-157.
* 5. *Didactics and Mathematics.* Palo Alto, California: Creative Publications, 1978.
 6. Denmark, Tom. "Reaction Paper: Classroom Diagnosis." In Jon Higgins and James Heddens (eds.), *Remedial Mathematics: Diagnostic and Prescriptive Approaches.* Columbus, Ohio: ERIC Clearinghouse for Science, Mathematics and Environmental Education (ERIC/SMEAC), 1976.
 7. Engelhardt, Jon M. "Remediation of Learning Difficulties in School Mathematics: Promising Procedures and Directions." In Jon Higgins and James Heddens (eds.), *Remedial Mathematics: Diagnostic and Prescriptive Approaches.* Columbus, Ohio: ERIC/SMEAC, 1976.
 8. Gow, Doris T. "A Synthesis of Research in Basic Skills." Pittsburgh, Pennsylvania: Learning Research and Development Center, University of Pittsburgh, 1977.
 9. Herold, Persis Joan. *The Math Teaching Handbook.* Newton, Massachusetts: Selective Educational Equipment, Inc., 1978.
 10. Inskeep, James E., Jr. "Diagnosing Computational Difficulty in the Classroom." In M. Suydam and R. Reys (eds.), *Developing Computational Skills,* 1978 NCTM Yearbook. Reston, Virginia: NCTM, 1978.
 11. Johnson, Stanley W. *Arithmetic and Learning Disabilities: Guidelines for Identification and Remediation.* Boston, Massachusetts: Allyn and Bacon, Inc., 1979.
 12. Junge, Charlotte W. "Adjustment of Instruction (Elementary School)." In *The Slow Learner in Mathematics,* 35th NCTM Yearbook. Reston, Virginia: NCTM, 1972.
 13. Lankford, Francis G., Jr. "What Can a Teacher Learn about a Pupil's Thinking through Oral Interviews?" *The Arithmetic Teacher,* 21, 1 (January 1974): 26-32.
*14. Payne, Joseph N. (ed.) *Mathematics Learning in Early Childhood,* 37th NCTM Yearbook. Reston, Virginia: NCTM, 1975.
*15. Reisman, Fredricka K. *A Guide to the Diagnostic Teaching of Arithmetic.* Columbus, Ohio: Charles E. Merrill, 1978.
 16. Riedesel, C. Alan. "Reaction Paper: Clinical Diagnosis of Children with Mathematical Difficulties." In Jon Higgins and James Heddens (eds.), *Remedial Mathematics: Diagnostic and*

References (continued)

    Prescriptive Approaches. Columbus, Ohio: ERIC/SMEAC, 1976.

*17. Suydam, Marilyn, and Robert Reys (eds.). *Developing Computational Skills.* 1978 NCTM Yearbook. Reston, Virginia: NCTM, 1978

18. Underhill, Robert G. "Classroom Diagnosis." In Jon Higgins and James Heddens (eds.), *Remedial Mathematics: Diagnostic and Prescriptive Approaches.* Columbus, Ohio: ERIC/SMEAC, 1976.

19. Weaver, J. Fred. "Big Dividends from Little Interviews." *The Arithmetic Teacher,* 22, 4 (April 1955): 40-47.

*The references with asterisks are readings we believe would be especially interesting to and useful for teachers.

Research Within Reach
Elementary School Mathematics

Unlocking the Mind of a Child: Teaching for Remediation in Mathematics

Question: I am a sixth-grade mathematics teacher. About a third of my students seem to be at the second-grade level in basic skills. How do I bring their skills up to an appropriate level in a way that will keep them motivated in the face of material they are seeing for the fourth or fifth year in a row?

Effective remediation begins with effective diagnosis. The sharp focus on individual needs and strengths that emerges from diagnostic activities should form the foundation of any remedial program. Where there is an absence of clarity about individual student needs, it is all the more tempting to clump children in categories such as "slow learners," "underachievers," or "remedial students."

None of these categories is ever defined with any degree of precision, but, generally, children are called *slow learners* in mathematics if both their achievement and their mathematical self-image are poor (sometimes, if their IQ is low as well), *underachievers* if their achievement fails to measure up to their potential, and *remedial students* if they work within a curriculum that is tailored to help them "catch up" in mathematics.

To the extent that it reminds teachers of the differences among learners, such categorizing is a useful tool. However, there is always the dangerous possibility that the categorizing will give way to labelling. The result, all too often, is the misreading of a child's potential for mathematical learning and the subsequent delay or even denial of certain mathematical learning experiences (12). In contrast, the richest opportunities for children arise in settings where there is close attention to *individual* needs and *individual* potential.

From Diagnosis to Remedial Goals

Diagnostic observation, testing, and interviews can uncover a variety of information about each child in such areas as

1. Thinking strategies, for example, she readily uses a number line to approach addition and subtraction exercises.
2. Systematic errors, for example, he consistently fails to regroup in subtraction.
3. Areas of strength, for example, she knows addition and subtraction facts up to 18.
4. Areas of confidence, for example, he is eager to participate in basic facts drill.
5. Developmental level, for example, she has difficulty working addition and subtraction exercises without the aid of a number line.
6. Motivational needs, for example, he seems easily distracted away from written exercises.

As you diagnose the child's strengths and needs, you have to gauge how far she is from mastery of the prerequisites for the current mathematical objectives in your class and, as a result, decide whether to plan a remedial program.

If there are but a few weak points, you can try some concentrated, short-term work in the child's mathematical trouble areas. For example, children who consistently make careless errors due to column misalignment can use graph paper to record their work; children who regularly add to solve subtraction exercises, because they ignore the subtraction sign, can be asked to circle the sign in each exercise before they embark on their computation.

Instead of this:

$$\begin{array}{r}147\\7\\\underline{93}\end{array}\qquad\begin{array}{r}46\\-35\\\hline 81\end{array}$$

Try this:

$$\begin{array}{r}1\,4\,7\\7\\\underline{9\,3}\end{array}\qquad\begin{array}{r}46\\\ominus 35\\\hline 11\end{array}$$

Many children will get trapped in habits of error in spite of an understanding of the workings of algorithms. For them the pitfalls may be remembering the steps of the algorithm or remembering basic facts. In the former instance, you can work with the child (or children) to make up a chart that represents the steps of the algorithm; in the latter case, you can organize some focused, short-term group practice in basic facts.

Obviously, short-term solutions will not always apply. If your diagnosis shows a child to be far removed from mastery of prerequisites, you should consider a more comprehensive remedial program. In the remainder of this report we present suggestions, derived from classroom investigations, for grouping students for remedial work, for setting remedial program goals, and for choosing appropriate teaching methods.

As your planning for a child's remedial program takes shape, involve him in the planning (1). Discuss with him the grouping procedure, your purposes in setting up the program, and the content areas that will be covered.

Grouping for Remediation

With the information provided by your diagnostic efforts to identify developmental levels in your class and missing prerequisites for each student, you can begin to group for remedial work.

- Be flexible with your grouping and design the groups to be temporary. Allow the sizes and the number of groups to vary with the content of the lesson and the developmental levels of the children (8).

- Keep appropriate materials on hand and let the children choose from the available materials (1). For example, if a group is working on division with one-digit numbers, the children should be able to check $248 \div 8$ concretely, with bundles of hundreds, tens, and ones, if that is the route most comfortable and meaningful for them.

- Encourage discussion within a group about discoveries they have made. One writer has pointed out that "the class sharing and discussion involved in group discovery may be one of the best ways to meet individual differences" (8).

- Form temporary student teams for practice on basic facts and for remedial help on computational skills. To insure the effectiveness of the practice, you can choose as one of the team members a child who is competent in the area being practiced (8).

Remedial Goals

Investigators studying remediation stress that you should deal with more than arithmetic in your design of remedial programs. Specifically, a remedial program should include instruction on how to locate information, how to use the textbook, how to study, how to remember, and how to check computation (7). In the same vein, studies also suggest that you:

- Provide consistent and clear guidance for the child in developing organizational skills (1)—how to arrange numbers in computation, how to organize data from word problems, and so on. Stress the value of visually neat and organized work.

- Aim to strengthen the child's self-image (1). There is convincing evidence that the child who is way behind her peers in mastery bears the burden of a low self-image, which can operate in a vicious cycle to keep her mired away from mastery (12). One suggested method of shoring up a child's self-image is to provide positive, immediate praise and reinforcement for success (1). And if a child, especially a low achieving child, receives that praise and acknowledgment of success in *public* situations, his self-image will benefit even more (2).

One vehicle for public acknowledgment is a progress chart for each student, which, incidentally, can aid in your planning of instruction. Further suggestions for developing self-image in students are in our *Research Within Reach* report entitled "Motivation in Mathematics."

- Encourage the development of good listening skills (12). If listening is a problem with a particular child, be an active listener yourself, and acknowledge questions and statements in a way that lets her know you are listening to her. You can also make it a habit to ask individual students to express in their own words whatever it is you are hoping they have heard. Of course, if you consistently single out the poor listeners for this, the whole experience will be viewed as punitive, so you must take care to involve all the students.

- When you are working with a child on basic facts, make sure that he understands the facts, and can show he understands them, before you expect memorization. Children should be able "to provide both physical and mathematical evidence that they understand an arithmetic fact before they are asked to memorize the corresponding symbolic statement" (3). For example, children give evidence of understanding the statement "$5 + 2 = 7$" if they can give responses similar to these:

> "Five plus two equals seven is true because if I stand on five on the number line and take two steps forward, I can get to seven."

> "Here are five fingers and here are two. Put them together and you have seven fingers."

Davis provides nine more principles for guiding the child, with drill, encouragement, and reinforcement, toward a mastery of basic facts:

1. Have children begin to memorize basic arithmetic facts soon after they demonstrate an understanding of symbolic statements.
2. Children should participate in drill with the intent to memorize.
3. During drill sessions, emphasize remembering—don't explain!
4. Keep drill sessions short and have some drill almost every day.
5. Try to memorize only a few facts in a given lesson and constantly review previously memorized facts.
6. Express confidence in your students' ability to memorize—encourage them to try memorizing and see how fast they can be.
7. Emphasize verbal drill exercises and provide feedback immediately.

8. Vary drill activities and be enthusiastic.
9. Praise students for good efforts—keep a record of their progress.

- Help the child (or group) doing remedial work to see some underlying unity in the mathematics she's learned and to be able to extend it to new situations. Junge (8) recommends, as one avenue, an emphasis on patterns—color patterns, shape patterns, number patterns, even sound patterns ("Close your eyes and listen to these drum taps. See if you can pick out the pattern.")—and lists a series of examples for all levels of elementary school, such as the following type of exercise:

$8 \times 7 = 56$, so $16 \times 7 = \underline{2 \times 8 \times 7 = 112}$
$9 \times 9 = 81$, so $18 \times 9 = \underline{\hphantom{xxxxxxxxxx}}$
$7 \times 5 = 35$, so $14 \times 5 = \underline{\hphantom{xxxxxxxxxx}}$
$6 \times 15 = 90$, so $12 \times 15 = \underline{\hphantom{xxxxxxxxxx}}$

Remedial Methods

Some suggested methods for effective remediation are:

- Design remedial instruction to be different from previous instruction (1). For example, if a child errs systematically at one or more stages of a subtraction algorithm, you should look to alternative ways to guide the child to the understanding and use of an algorithm for subtraction.

For some students a step-by-step progression, from a concrete embodiment of the algorithm to its symbolic representation, will be meaningful. (See reference #9 for such a progression). Others may be helped by an expanded version of the same algorithm or by an alternative algorithm (1 and 5). An example might be

$$\begin{array}{r} & & 110 \\ & 500 & \cancel{70} & 15 \\ 625 = & \cancel{600} + \cancel{20} + \cancel{5} \\ -348 = & 300 + 40 + 8 \\ \hline & 200 + 70 + 7 = 277 \end{array}$$

- There is research evidence that some children think in more spatial terms than others. For that reason, as well as for motivational and developmental reasons, multisensory experiences are important. Researchers (1, 8, 13) recommend that you incorporate concrete and pictorial approaches throughout your remedial repertoire. Use them in conjunction with the symbolic approaches at each step of a child's remedial program. For example, if the topic is renaming equivalent fractions, the three-step procedure recommended is as follows (illustrations are from reference #11):

  1. Concrete (also called enactive) level: have the child compare equivalent regions of identical wholes, using cutouts from rectangular pieces of paper and leaving one whole, cutting a second in halves, a third in fourths, and a fourth in eighths. Guide the child to place four eighths over ½ and two fourths over ½.

  2. Pictorial (also called iconic) level: have the child compare equivalent regions

  $$\frac{1}{2} = \frac{2}{4} = \frac{4}{8}$$

  3. Symbolic level: use the concept of multiplicative identity

  $$\frac{1}{2} \times 1 = \frac{}{4} \qquad \frac{1}{2} \times \frac{2}{2} = \frac{2}{4}$$

- Several researchers (4, 8) have stressed the importance of your gradually guiding the children from a concrete, intuitive understanding of mathematical principles toward the ability to represent their understanding verbally and symbolically. Merseth's article (9) sets down a program in which a child can comfortably and systematically transform her intuitive perceptions into verbal and symbolic expressions.

- Encourage the child to estimate answers (1), both as a preparation for checking computational accuracy and as a way to develop problem solving skills. For further information, see other reports in the Research Within Reach series, "Mathematical Problem Solving: Not Just a Matter of Words" and "Estimation and Mental Arithmetic."

- Use hand-held calculators in your remedial instruction. Though the value of calculators to the mathematics curriculum goes far beyond remediation, students doing remedial work can benefit greatly by using them to sharpen their estimation and problem-solving skills.

There is no reason to assume that mathematical comprehension, problem-solving skills, or an understanding of applications are dependent on a mastery of computation (10). For a student temporarily off course in the learning of computational skills, a calculator can provide the freedom and safety to develop other mathematical skills. Examples (from reference #6) follow:

1. Give the students a square array of numbers, for example,

18	26	12
25	33	19
55	63	49

Inform them that certain groups of addends from the square will equal a target number, for example, 100.

Ask them to use their calculators to try and find two such groups.

2. You can ask the children to use their calculators to answer questions:

How old are you in months?
How old are you in weeks?
How old are you in days?

Conclusion

An effective remediation program demands careful planning. The above suggestions are some of the guidelines studies have associated with appropriate remedial goals and successful remedial methods. No matter how your class is organized for mathematics instruction, those guidelines can fit comfortably with your teaching, provided you are committed to working within the framework of each child's needs and to maintaining an attitude of flexibility as you continue to deal with those needs.

The rewards can be rich. Several studies have indicated that, given sufficient time and appropriate types of help, more than 90 percent of students can learn a subject with a high degree of mastery (10). Other studies are not so astounding in their figures, but *all* point to increased achievement and motivation when remedial programs are carefully designed.

References

* 1. Ashlock, R.B. *Error Patterns in Computation.* Columbus, Ohio: Charles E. Merrill, 1972.

2. Brophy, Jere E., and Good, Thomas L. *Teacher-Student Relationships: Causes and Consequences.* New York: Holt, Rinehart & Winston, 1974.

3. Davis, Edward J. "Suggestions for Teaching the Basic Facts of Arithmetic." In M. Suydam and R. Reys (eds.), *Developing Computational Skills,* 1978 NCTM Yearbook. Reston, Virginia: National Council of Teachers of Mathematics, (NCTM) 1978.

4. Engelhardt, Jon M. "Remediation of Learning Difficulties in School Mathematics: Promising Procedures and Directions." In Jon Higgins and James Heddens (eds.), *Remedial Mathematics: Diagnostic and Prescriptive Approaches.* Columbus, Ohio: ERIC Clearinghouse for Science, Mathematics, and Environmental Education, 1976.

5. Hutchings, Barton. "Low-Stress Algorithms." In Doyal Nelson and Robert Reys (eds.), *Measurement in School Mathematics,* 1976 NCTM Yearbook. Reston, Virginia: NCTM, 1976.

6. Immerzeel, George, and Earl Ockenga. *Calculator Activities for the Classroom.* Palo Alto, California: Creative Publications, Inc., 1977.

7. Inskeep, James E., Jr. "Diagnosing Computational Difficulty in the Classroom." In M. Suydam and R. Reys (eds.), *Developing Computational Skills,* 1978 NCTM Yearbook. Reston, Virginia: NCTM, 1978.

8. Junge, Charlotte W. "Adjustment of Instruction (Elementary School)." In *The Slow Learner in Mathematics,* 35th NCTM Yearbook. Reston, Virginia: NCTM, 1972.

* 9. Merseth, Katherine Klippert. "Using Materials and Activities in Teaching Addition and Subtraction in Algorithms." In M. Suydam and

References (continued)

R. Reys (eds.), *Developing Computational Skills*, 1978 NCTM Yearbook. Reston, Virginia: NCTM, 1978.

10. Pikaart, Len, and James W. Wilson. "The Research Literature." In *The Slow Learner in Mathematics*, 35th NCTM Yearbook. Reston, Virginia: NCTM, 1972.

*11. Reisman, Fredricka K. *A Guide to the Diagnostic Teaching of Arithmetic*. Columbus, Ohio: Charles E. Merrill, 1978.

12. Schulz, Richard W. "Characteristics and Needs of the Slow Learner." In *The Slow Learner in Mathematics*, 35th NCTM Yearbook. Reston, Virginia: NCTM, 1972.

*13. Suydam, Marilyn N., and Jon L. Higgins. *Activity-Based Learning in Elementary School Mathematics: Recommendations from Research*. Reston, Virginia: NCTM, 1977.

---

*The references with asterisks are readings we believe would be especially interesting to and useful for teachers.

Research Within Reach
Elementary School Mathematics

Evaluation in Mathematics Education
Part One: Looking Beneath and Beyond the Tests

Question: I am a third-grade teacher. Our mathematics program uses pretests to place students into certain units, but I find many students are misplaced. My students often miss things on a pretest that they actually know. Is there a better way to test?

Tests are not infallible; neither are they educational ends in themselves. To the extent that this is accepted, teachers can do a lot to make tests work in their best interests. To the extent that they are willing to look beneath test results with a mildly skeptical and critical eye, and to look beyond tests to the goal of improved instruction, teachers can make tests work in the best interests of their students, too.

Toward those ends, teachers need to adopt a wider perspective toward classroom evaluation than one which points to testing alone as the method of evaluating, or to grading alone as the means of reporting.

Here is a brief, informal glossary of evaluation terms:

1. If the purpose of the evaluation is to compare the student with peers (through percentile, grade equivalent scores, and so on), then it is a *norm-referenced* evaluation.

2. If the evaluation is made on the basis of a predetermined standard, and not on the basis of peer comparison, the evaluation is said to be *criterion-referenced* (for example, Boy Scout and Girl Scout promotions are based on criterion-referenced evaluations).

3. If the evaluation is made at the end of a period of instruction, with no eye toward revising future instruction, but only to determine what has been learned during the period, it is called *summative* evaluation.

4. If the evaluation is made at the end of a period of instruction and is designed specifically so that the future instruction of the child

can be revised on the basis of the evaluation, it is called *formative* evaluation.

Evaluation Goals and Objectives

The roots of evaluation lie in the setting of educational goals and objectives, a process that often operates at some distance from the classroom teacher. From state, district, school, and publisher, or some combination thereof, comes a collection of standards for judging classroom learning, and those objectives invite the most convenient form of evaluation, the paper and pencil test. However, it is important for teachers to look beyond the appeal of convenience and to weigh the importance of educational goals that may not always lend themselves to paper and pencil evaluation. Following are several examples of these broader goals:

- We note in the *Research Within Reach* bulletin, "Mathematical Problem Solving: Not Just a Matter of Words," that problem solving skill is nourished by a number of related skills. For example, the flexibility to move from an unsuccessful approach to a problem to another approach is a hallmark of a good problem solver. We also note in that bulletin that it is possible to build the development of problem solving flexibility into instruction—by requiring students to list a number of possible approaches before they lock themselves into one that may or may not be fruitful. Once this has become a conscious teacher goal, then each child can be evaluated in light of her responsiveness, or lack of responsiveness, to this aspect of instruction. As an aid in keeping himself conscious of such goals, the teacher can maintain a brief checklist for each student. A checklist for problem solving might begin in the following way:

	Never	Sometimes	Regularly
1. Suggests route to solution.			
2. Checks to see if answer is reasonable.			
3. Makes second attempt			
4. ...			

As a result of attending to this checklist, the teacher can make appropriate instructional adjustments so that a child might, for example, show more problem solving flexibility in the future.

- Long-term retention of mathematical learning, as well as quick and efficient relearning, should be goals of all teachers for their students. As goals, they should influence teachers' planning and evaluation and should lead to appropriate instructional changes. (See the *Research Within Reach* bulletin "Meaning in Elementary School Mathematics" for more information on retention.)

- Motivation, mathematical self-concept, and the quality of peer interactions are all important aspects of a child's school experience that invite goal setting and evaluation as much as mathematical content does (see the *Research Within Reach* bulletin "Motivation in Mathematics").

These examples are cited to make the point that teachers need to consider the entire spectrum of a child's school experience in planning for evaluation only the teachers themselves can carry out. Other examples can be found in references 3 and 12.

Evaluation by Paper and Pencil Tests

The most familiar form of evaluation, of course, is the paper and pencil test, either teacher-made or commercially published (standardized). Testing is done in a variety of ways, with a variety of labels and objectives. Achievement tests, aptitude tests, mastery tests, competency tests, and diagnostic tests are several types teachers frequently encounter.

Many tests are norm-referenced, some are criterion-referenced, and some are a blend of the two. For example, the test may be broken down into mathematical objectives for mastery in specific areas, but in the end the children are compared, by score, with their peers.

Both the writing of tests by teachers and the selection of standardized tests must be carried out with care. In *Evaluation in the Mathematics Classroom* (12), Suydam provides specific guidelines for both. For the former, there are tips about writing style, cautionary comments about test distractors, visual and organizational pointers, and a number of other essential components of good test construction.

For the selection of standardized tests the guidelines stress the importance of examining the tests to ensure that they match with teacher purpose and student experience. A recent study by Floden et al. (4) underscores the fact that this should not be taken lightly. Floden and his colleagues analyzed the four most popular mathematics achievement tests for the fourth grade (Iowa Tests of Basic Skills, Stanford Tests of Basic Skills, CTB/McGraw-Hill Comprehensive Tests of Basic Skills, and Metropolitan Achievement Tests) and determined that they clearly do not measure achievement in the same content areas. For example, one test presents more than 40 percent of its items in the form of graphs, figures, and tables, while another presents only 15 percent of its items in this way. One test contains six items testing percent knowledge, while two of the others have no percent items at all. Throughout the thirty categories considered by the researchers in this study, such discrepancies appeared frequently.

- In the specific area of diagnostic testing, interested teachers can profit from reading the *Research Within Reach* bulletin, "Diagnosis: Taking the Mathematical Pulse," and the works of Reisman (9) and Johnson (8), in addition to Suydam's guide.

Evaluation at the Primary Level

No method of evaluation is foolproof. While all teachers need to stay alert to that fact, it is especially important at the primary level. Teachers of very young children should always question whether the behavioral indications of *mastery* (for example, the connecting of dots between two columns) accurately reflect conceptual *understanding* (in this example, one-to-one correspondence). In fact, it is unclear whether one can talk about mathematical mastery for children in primary grades in the same way one talks about it for older children.

It is clear that young children are limited in mathematical mastery by cognitive development (see the *Research Within Reach* bulletins, "Mathematics in Kindergarten" and "The Bridge from Concrete to Abstract"). What needs attention here is the effect those limitations have on evaluation.

First of all, there is every indication that the younger the child is, the more literal and concrete should be our interpretation of the Piagetian aphorism, "to know is to act on an object." Manipulation of objects is an integral part of a young child's learning, and his level of cognitive development casts a definite shadow over any paper and pencil attempt to gauge how well a particular learning has been absorbed—for example, one-to-one correspondence.

Compounding evaluation even further is the fact that the young child is easily *distracted by irrelevant detail*. In one study, reported by Stevenson (11), researchers presented third and fourth graders with the task of choosing from groups of four stimuli the one in each group which differed from the others in color. Some of the items contained irrelevant information—differences in size, brightness, or thickness—and the children not only fell victim to the distractions initially, but persisted in being distracted even after training. (See also Bana and Nelson [1].)

Language also presents a potential barrier to evaluating mathematics learning. For example, would a child's interpretation of the following test item be the same as the test designer's?

How many objects are there altogether?

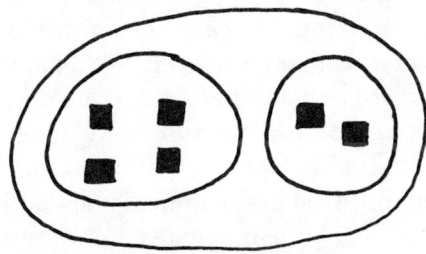

Some children will respond "6"; others might very well respond "9."

Another factor relevant to evaluation is a child's *depth of conviction* as she develops from one cognitive level to another. For example, around the age of seven or eight she may consistently and accurately use the concept of one-to-one correspondence, but it is often possible to shake her confidence with some manipulation of objects (for example, bunching up one of two equivalent groups of checkers) (6).

Teachers of young children, then, need to stay mindful of distractors, confusing language, and each child's depth of conviction when they assess mastery in the primary grades. Some helpful suggestions emerge from research and practice:

- Ask each child, in one-on-one interviews, to put into her own words what she thinks you are asking her to do and what key words mean to her. From his study with entering first graders, Hendrickson concluded that much more of this kind of interaction needs to occur (7).

- Present problems so that critical verbal and visual features are highlighted (11).

- As often as possible, use problem settings where the child has room for a variety of responses. The 37th Yearbook of the National Council of Teachers of Mathematics (NCTM) offers many examples. The following one is in the area of one-to-one correspondence:

    Show a model set of objects. Have the children sort other sets into those that are equivalent to the model set and those that are not. Ask how they can tell. (5)

While these suggestions need attention, it is significant that a number of mathematics educators urge a tempered tone toward the entire issue of mastery at the primary level. For example, several educators who have had considerable success in developing measurement concepts ("longer," "taller," "heavier," and so on) in kindergarten and first grade emphasize that it is essential that teachers not expect children to *master techniques* (such as finding out how to read a ruler). Rather, teachers should strive to *make comparisons meaningful*. Thus, children may not be able to measure two buildings with a ruler, but they can talk about one as taller (10).

Apparently, similar conclusions can be made about mastery in all areas of the young child's learning. After their extensive study of teacher expectations in the classroom, Brophy and Good were led to write:

> Such observations have led us to conclude that too much emphasis in schools is placed upon mastering the curriculum, *especially in early grades*, and too little emphasis has been placed upon helping students to develop appropriate learning sets. Of prime importance is that students learn to expect that they can achieve, to approach the teacher and to ask a question when they do not understand an explanation or an assignment, to understand that the classroom is a place where they can find answers to their own questions at times, and to understand that evaluation is simply feedback guiding future efforts, not personal ratings of one's value. (2)

This brings us full circle to the beginning message of this report—tests are but one means of evaluating students, and teachers must not accept test results at face value. There are too many factors involved in a child's learning growth for all to be captured on a test, no matter what the nature of that test. At the core of effective and fair evaluation is a wide range of teacher goals and objectives, and a consistent awareness, through observation, interview, and testing, of each child's performance in the light of those goals and objectives.

References

1. Bana, Jack, and L. Doyal Nelson. "Distractors in Nonverbal Mathematical Problems." *Journal for Research in Mathematics Education.* 9 (January 1978): 55-61.

2. Brophy, Jere E., and Thomas L. Good. *Teacher-Student Relationships: Causes and Consequences.* New York: Holt, Rinehart and Winston, Inc., 1974.

* 3. *Didactics and Mathematics.* Palo Alto, California: Creative Publications, 1978.

References (continued)

4. Floden, Robert E.; Andrew C. Porter; William H. Schmidt; and Donald J. Freeman. *Don't They All Measure the Same Thing? Consequences of Selecting Standardized Tests.* East Lansing, Michigan: Institute for Research on Teaching, 1978.

* 5. Gibb, E. Glenadine, and Alberta M. Castaneda. "Experiences for Young Children." In *Mathematics Learning in Early Childhood*, 37th NCTM Yearbook. Reston, Virginia: National Council of Teachers of Mathematics (NCTM), 1975.

6. Ginsburg, Herbert. "Young Children's Informal Knowledge of Mathematics." *Journal of Children's Mathematical Behavior*, 1 (Summer 1975): 63-156.

7. Hendrickson, A. Dean. "An Inventory of Mathematical Thinking Done by Incoming First Grade Children." *Journal for Research in Mathematical Education*, 10 (January 1979): 7-23.

8. Johnson, Stanley W. *Arithmetic and Learning Disabilities: Guidelines for Identification and Remediation.* Boston, Massachusetts: Allyn and Bacon, Inc., 1979.

* 9. Reisman, Fredricka K. *A Guide to the Diagnostic Teaching of Arithmetic.* Columbus, Ohio: Charles E. Merrill, 1978.

10. Robinson, G. Edith; Michael I. Mahaffey; and L. Doyal Nelson. "Measurement." In *Mathematics Learning in Early Childhood*, 37th NCTM Yearbook. Reston, Virginia: NCTM, 1975.

11. Stevenson, Harold W. "Learning and Cognition." In *Mathematics Learning in Early Childhood*, 37th NCTM Yearbook. Reston, Virginia: NCTM, 1975.

*12. Suydam, Marilyn N. *Evaluation in the Mathematics Classroom.* Reston, Virginia: National Council of Teachers of Mathematics and ERIC Clearinghouse for Science, Mathematics, and Environmental Education, 1974.

*The references with asterisks are readings we believe would be especially interesting to and useful for teachers.

Research Within Reach
Elementary School Mathematics

Evaluation in Mathematics Education
Part Two: Mastery Learning in Elementary School Mathematics

Question: I am a third-grade teacher. Our mathematics program uses pretests to place students into certain units, but I find many students are misplaced. My students often miss things on a pretest that they actually know. Is there a better way to test?

In Part One, "Looking Beneath and Beyond the Tests," we talked in general terms about evaluation in the mathematics classroom, pointing to the importance of looking beyond mere paper and pencil testing. We stressed the need for teachers to do *formative* evaluation—that is, to use testing and other forms of evaluating a child's understanding and mastery of mathematics in order to adjust future instruction.

In this bulletin we highlight the mastery learning model, drawn from a theory which has at its core frequent and systematic formative evaluation. The model's proponents have made certain assumptions about learning and have planned strategies on the basis of those assumptions.

Some Mastery Learning Assumptions

A very large percentage of students can master the school curriculum. Advocates of mastery learning suggest that a student's aptitude be thought of as the amount of time needed for learning when the quality of instruction best suits that individual's needs and she is persevering. Bloom (2) claims that where such a perspective prevails, 90 percent of all students can learn at a level now reserved for the top 10 percent.

There are five variables involved in learning for mastery.

1. Aptitude for particular types of learning, viewed as the amount of time required by the learner to attain mastery of the task.

2. Quality of instruction, viewed in terms of its approaching the optimum for a given learner.

3. Ability to understand instruction, that is,

to understand the nature of the task and the procedures to follow.

4. Perseverance, the amount of time one is willing to spend in learning.

5. Time allowed for mastery. (1)

Educational failure is closely tied to a reliance on norm-referenced testing. Mastery learning proponents point out that norm-referenced tests, *by design*, put a certain percentage of students into the category of failure. They are particularly disturbed by the research evidence that sixth-grade achievement levels in several areas are highly predictable four years earlier on the basis of standardized test scores. In other words, once a second-grade student is deemed a failure on the basis of standardized scores, it is extremely difficult for the student to break away from that label, even by the sixth grade.

Formative evaluation should be done regularly and frequently, after small blocks of learning. One of the cornerstones of the theory is a commitment to mastery of *clearly specified* learning objectives (for example, multiplication of three-digit numbers) soon after the objectives are set. Through frequent evaluation, proponents believe, instruction becomes more efficient and learning accumulates faster.

Mastery Learning Strategies

The learning objectives and curricular details may vary from one mastery learning classroom to another, but the basic strategy seems to remain simple and constant. A course is arranged as a sequence of learning units and the following 5-step procedure is repeated for each such learning unit (8):

The earliest units are earmarked for special attention, on the theory that thorough mastery of early units will make later learning more effective and efficient. Thus, smaller units and more instructional time may be the rule toward the beginning of a school year, but the foundation that results is seen as one on which students can readily build.

While all mastery models draw their life from close and continued attention to this 5-step procedure, there is some variation in classroom organization. In fact, mastery programs seem to fall into two distinct camps, the one committed to group instruction, the other to individualized instruction (10).

Research on Mastery Learning

Earlier we noted the roles played in mastery theory by perseverance and the ability to understand the nature of learning tasks and procedures. Perhaps because of these relatively sophisticated expectations, very few of the reported applications of the mastery model are found at the primary level. Primary teachers should approach the issue of mastery with some care, because, as we noted in Part One, there are developmental factors which may color any judgments about mathematical mastery for very young children.

There have been some applications and research at the intermediate and junior high level, and perhaps all teachers can draw wisdom from the results:

What is a reasonable definition of mastery? Here the words of Block (1) best sum up the relevant research:

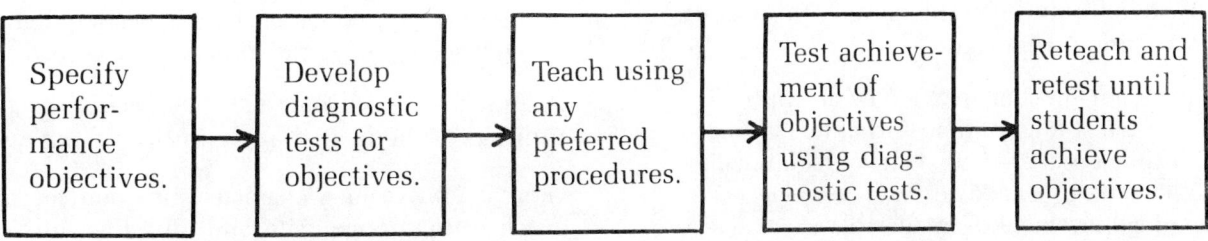

> The empirical work to date suggests that if students learn 80 to 85 percent of the skills in each unit, then they are likely to exhibit maximal positive cognitive and affective development as measured at the subject's completion. This work also suggests that encouraging or requiring students to learn all or nearly all (90 to 95 percent) of each unit, besides being an unrealistic expectation in terms of student and teacher time and effort, . . . may have marked negative consequences for student interest in and attitudes toward the learning. (p. 70)

One result of these findings is that in many mastery learning programs five items for each learning objective appear on achievement tests, and students are assumed to have achieved mastery of the objective if they have at least four of the five correct.

Teachers who adhere too rigidly to these practices are courting danger, however, if they teach only those concepts and skills their students are ready to master, that is, the concepts and skills for which they have mastered the prerequisites. For example, it is possible, even *advisable*, to plant the seeds of concepts like angle, shape, and length long before a child is ready to work with geometry at a mastery level (6).

How well does the mastery learning model work? From a broad look at the literature on mastery learning, Bloom (2) concluded that

> there is considerable evidence that mastery learning procedures do work well in enabling about four-fifths of students to reach a level of achievement which less than one-fifth attain under non-mastery conditions. (p. 5)

Mastery theorists cite evidence that students become more efficient in their learning as they proceed through a mastery program. After reviewing several studies on time use in mastery programs, Bloom reported:

> Perhaps the main conclusion to be drawn from these limited studies of time—whether elapsed time or time-on-task—is that rate of learning or amount of time needed to learn to some criterion of achievement is an alterable characteristic. That is, when students are provided with the time and help they need to learn and when this produces positive entry characteristics (cognitive and affective),* students not only become better able to learn, they also become able to learn with less and less time. (p. 191)

Drawing this line of thinking a bit further, mastery proponents maintain that learning how to learn develops a momentum of its own through the mastery model.

Critics, however, are not so quick to accept that it is learning how to learn which fuels the momentum. In fact, Erlwanger (5) concluded from his case studies that at least some children in a mastery program can appear by test scores to have mastered fifth-grade mathematics, while in actuality they have mastered *skills and procedures* with very porous *understanding*. There is the possibility, he concludes, that mastery models in mathematics produce learners who are adept at manipulating marks on paper while they lack the skill to tie together mathematical concepts.

How extensive are time and energy investments for teachers wanting to adopt a mastery learning model? In addressing this issue, so critical for all teachers, we point first to a research conclusion which is rather broad in scope, and then to a study which focused on the training of teachers for the mastery learning classroom.

When Bloom (2) discussed the apparent success of mastery learning with four-fifths of all students exposed to these procedures (see our quotation above), he also offered the following observation:

> The time costs for this are typically of the order of 10 to 20 percent additional time over the classroom scheduled time. The efficiency of the correctives and the additional time needed are direct functions of the diagnostic-progress feedback testing—the formative tests. (p. 5)

Bloom believes that, as the teacher becomes more proficient at diagnosing the areas where

---

*For "cognitive entry characteristic," one can substitute learning prerequisite, and "affective entry characteristic" refers to motivation.

each student falls short of mastery, less additional time is needed to make the mastery model work.

In a study by Okey, a group of teachers (kindergarten through eighth grade) was trained for less than five hours in the use of mastery teaching. They then employed the approach in half of their science and mathematics classes. Their training came from the *Teaching for Mastery* program designed by Okey and Ciesla, and with its help the teachers were able to produce somewhat higher achievement when they used the mastery teaching skills than when they did not (8).

On the negative side, strictly individualized programs have come under some criticism for the time demands they place on teachers and for their relatively high cost when compared with traditional programs. In his survey of the research on self-paced programs in mathematics, Schoen concluded that "self-paced instruction of the type used in these studies is very expensive to adopt" and "more work is expected of the teachers" than in traditional programs. He also pointed with concern to the evidence in several studies that "average and below-average students are likely to achieve less in these programs" (9).

Using a mastery approach, how quickly can students improve their rate of achievement? This question does not invite any single answer, as the following report of the results of two relevant studies shows.

In one study, a teacher who adopted a mastery approach compared the achievement in his fifth-grade arithmetic classes with the achievement of his non-mastery fifth graders the year before. The results indicated that there were significant increases in mastery across the board for the students exposed to the mastery approach. Of some note was the fact that one socio-economically disadvantaged class showed zero percent mastery on an arithmetic achievement test in the first year, and its counterpart in the following (mastery) year showed 20 percent mastery on the same test—progress nowhere near the 80 percent mentioned by Bloom, but a promising measure of progress nonetheless (7).

In a second study, teachers used a mastery learning strategy with elementary school students for a geometry unit. The data indicated that low mathematics aptitude fourth graders who were taught using a mastery strategy achieved at the same level as high aptitude fifth graders taught in a more conventional manner (3).

Conclusion

Mastery learning is based on a firm belief that almost every child can master the elementary school curriculum. As we have suggested in this bulletin, there is evidence that, at least insofar as *mathematical skills and procedures* are concerned, the belief holds up well in some applications of mastery theory. The doubts inspired by the theory's applications revolve primarily around the quality of the learning of *mathematical concepts and problem-solving techniques* and the relatively high cost incurred in individualized mastery programs.

Those teachers who wish to investigate formal mastery models in more depth should find the books and articles listed at the end of this bulletin helpful.

Others, less inclined to overhaul their classroom instruction with a formal model, can still benefit from this bulletin and the work of mastery learning theorists. To do so, they need only recognize the potential for improved achievement that accompanies *frequent and consistent* use of diagnosis, setting of objectives, use of formative testing, and adjustment of instruction.

References

* 1. Block, James H. (Ed.). *Mastery Learning: Theory and Practice.* New York: Holt, Rinehart, and Winston, 1970.
  2. Bloom, Benjamin S. *Human Characteristics and School Learning.* New York: McGraw-Hill, 1976.

References (continued)

3. Burrows, Charles K., and James R. Okey. *The Effects of Mastery Learning Model on Achievement.* Paper presented at the annual meeting of the American Educational Research Association, Washington, D.C., 1975. ED 109 240.

* 4. Crosswhite, F. Joe, and Robert E. Reys (Eds.). *Organizing for Mathematics Instruction,* 1977 NCTM Yearbook. Reston, Virginia: National Council of Teachers of Mathematics (NCTM), 1977.

5. Erlwanger, Stanley H. "Case Studies of Children's Conceptions of Mathematics—Part I." *The Journal of Children's Mathematical Behavior,* 1 (Summer 1975): 157-283.

* 6. Gibb, E. Glenadine, and Alberta M. Castaneda. "Experiences for Young Children." In *Mathematics Learning in Early Childhood,* 37th NCTM Yearbook. Reston, Virginia: NCTM, 1975.

7. Kersh, Mildred E. "A Strategy for Mastery Learning in Fifth-Grade Arithmetic." Unpublished Ph.D. dissertation, University of Chicago, 1970. (Abstracted in Block's *Mastery Learning: Theory and Practice.*)

8. Okey, James R. "Altering Teacher and Pupil Behavior with Mastery Teaching." *School Science and Mathematics,* 74 (October 1974): 530-535.

9. Schoen, Harold L. "Self-Paced Mathematics Instruction: How Effective Has It Been?" *The Arithmetic Teacher,* 23, 2 (1976): 90-96.

*10. Smith, Jeffrey K. *Perspectives on Mastery Learning and Mastery Testing.* ERIC/TM Report 63. Princeton, New Jersey: ERIC Clearinghouse on Tests, Measurement and Evaluation, 1977.

*The references with asterisks are readings we believe would be especially interesting to and useful for teachers.

Research Within Reach
Elementary School Mathematics

Motivation in Mathematics

Question: I am a sixth-grade mathematics teacher, and I need help in approaching the remedial needs of some of my unmotivated students, who are three years behind in skills. How can I motivate them to be interested in catching up while at the same time motivating the more successful students to work at their full potential?

Lighting fires of interest in mathematics and getting students to take strides for themselves—these are familiar tasks for many teachers.

There are no shortcuts to motivation, no quick formulas or programmed inducements. There are, however, some suggestions from research and from the experiences of teachers who have had some success in motivation. In this report we present some of these suggestions, with an eye toward the elementary school mathematics classroom.

First, however, a cautionary note is in order. At times the attention given to motivation may be misguided, while at other times it may be altogether insufficient. If Greg, for example, is disruptive or seems to forget basic mathematical facts too easily, his problem may be different from—and even more serious than—a lack of motivation. He may be learning disabled or developmentally immature. On the other hand, Marie may be regular in homework, cooperative in class, and willing to contribute her computational answers, all in apparent stark contrast with Greg. And yet Marie's need for increased and enriched motivation—for example, to improve her estimation and problem solving skills—could be just as great as that of any student in the class. Without such motivation, her interest in mathematics may begin to evaporate very quickly in junior high school, as indeed it *does* for many girls (6).

The suggestions we present are grouped under two headings, reflecting two separate, though related, vehicles for creating a motivational environment in the mathematics classroom: approaches to mathematical learning; and teacher-student communication.

Approaches to Mathematical Learning

- Enrich your mathematics curriculum by weaving in *regular* work with estimation strategies, problem solving, manipulatives, challenging drills, and so on. Each of these topics is the subject of a separate RDIS report in this series, so it suffices here to refer you to those reports and their references for the large number of enrichment suggestions they contain. One research recommendation is worthy of note, however, namely that teachers can and should expect students to enjoy the learning of mathematics and should make that expectation *evident* (3).

  Besides the curriculum enrichment just noted, one recommended way to make the expectation evident is to prepare and systematically use questions that are an invitation to think, not just tests of memory. Thus, while "How many 9's in 81?" has a place in your repertoire, so should questions like, "Can you see a pattern in those numbers? What do you think the next number will be?" It is important that children learn that teachers are interested in what they have to say and are not just hunting for the right answers (3).

- In a similar vein, encourage student-initiated questions. The value of this teaching strategy was confirmed in two studies that measured its effect on student achievement (1). From one of the same studies comes the further recommendation that you accept questions in the form they are asked. You may, for example, see some value in translating a question into more precise mathematical language, but it is apparently more important for the child to experience your acceptance first. The translation can come later, after you have made it clear that you value her asking the question.

Teacher-Student Communication

In this section we consider teacher-student communication which is less connected to the body of the mathematics curriculum and to the flow of mathematical questions and answers, but is still vital to student confidence, interest, self-concept, and persistence, that is, to student motivation.

- Work with each student, individually, to establish realistic, short-term goals (7). One motivation project showed that spending ten minutes a month talking with each individual child about his goals can produce dramatic results in learning (8). Such goals can run the gamut from those which are incidental to learning—for example, if the child would like to start being active in class, encourage her to speak up once each class —to specific learning objectives—for example, aspiring to mastery of addition with three-digit numbers.

  State clearly during each of those sessions that you expect a commitment on the part of the student to strive for the goal(s). The motivational importance of such statements has been demonstrated (3).

  For your part, if you seek commitments from students, it is vital that you also:

- Make clear to each student, especially the less successful students, that you will be supportive of them through failure situations. For example, your consistent waiting for responses and helping to draw out responses before you move on from one student to another is apparently important in the development of a student's confidence (1).

- Conduct regular (daily, if possible) and careful evaluations of each student's progress toward his goals, and keep the student aware of the progress (9). In one recommended procedure the teacher and student maintain a progress chart, record on it daily, and check it regularly (4).

- Provide a climate in your classroom in which risk-taking is perceived as acceptable. One characteristic of a low self-concept is a reluctance to take risks (2), and you can provide experiences in which risk-taking is both safe and interesting. For example, you can encourage educated guessing. ("How big a number do you think the answer will be?" "Can you guess a rule for

Instructional Strategies

this?") Allow a number of guesses each time you suggest guessing, record them all, and give them comparable consideration in the final determination of the answer. Both you and the students can learn much from incorrect answers.

- Be liberal and regular with positive reinforcement for risk-taking, thoughtful guessing, and perseverance. Several studies have recognized the importance of *public* acknowledgment of success, especially for low-achieving students as they experience success or progress in group situations (1, 3).

- Finally, as the child's confidence and self-concept begin to grow, it is important that you not only acknowledge success but also encourage her to take credit for her successes (1).

Conclusion

Several extensive studies have revealed that students frequently discern unspoken teacher expectations, and that those unspoken expectations can mold both teacher and student behavior in the classroom (2, 3). If the expectations are inappropriate (for example, "Charles's hesitation in answering always creates tension in the class"), then the resulting behavior will quite often stand in the way of student motivation (Charles is effectively denied the opportunity for public participation).

The teacher's potential for affecting student motivation, positively or negatively, is awesome. While, in the end, motivation must come from within each student, it can only come when the student feels the excitement of learning, experiences his or her efforts as appreciated, gets some clarity on goals, makes some connection between the work done in mathematics class and those goals, and feels the confidence and freedom to risk attaining them.

References

1. Borich, Gary D. "Implications for Developing Teacher Competencies for Process-Product Research." *Journal of Teacher Education*, 30, 1 (January-February, 1979): 79-85.

2. Braun, Carl. "Teacher Expectation: Socio-Psychological Dynamics." *Review of Educational Research*, 46, 2 (Spring 1976): 185-214.

3. Brophy, Jere E., and Thomas Good. *Teacher-Student Relationships: Causes and Consequences.* New York: Holt, Rinehart & Winston, 1974.

* 4. Davis, Edward J. "Suggestions for Teaching the Basic Facts of Arithmetic. In M. Suydam and R. Reys (eds.), *Developing Computational Skills*, 1978 NCTM Yearbook. Reston, VA: National Council of Teachers of Mathematics, 1978.

5. Engelhardt, Jon M. "Remediation of Learning Difficulties in School Mathematics: Promising Procedures and Directions." In Jon Higgins and James Heddens (eds.), *Remedial Mathematics: Diagnostic and Prescriptive Approaches.* Columbus, Ohio: ERIC Clearinghouse for Science, Mathematics and Environmental Education (ERIC/SMEAC), 1976.

6. Fox, Lynn H.; Elizabeth Fennema; and Julia Sherman. *Women and Mathematics: Research*

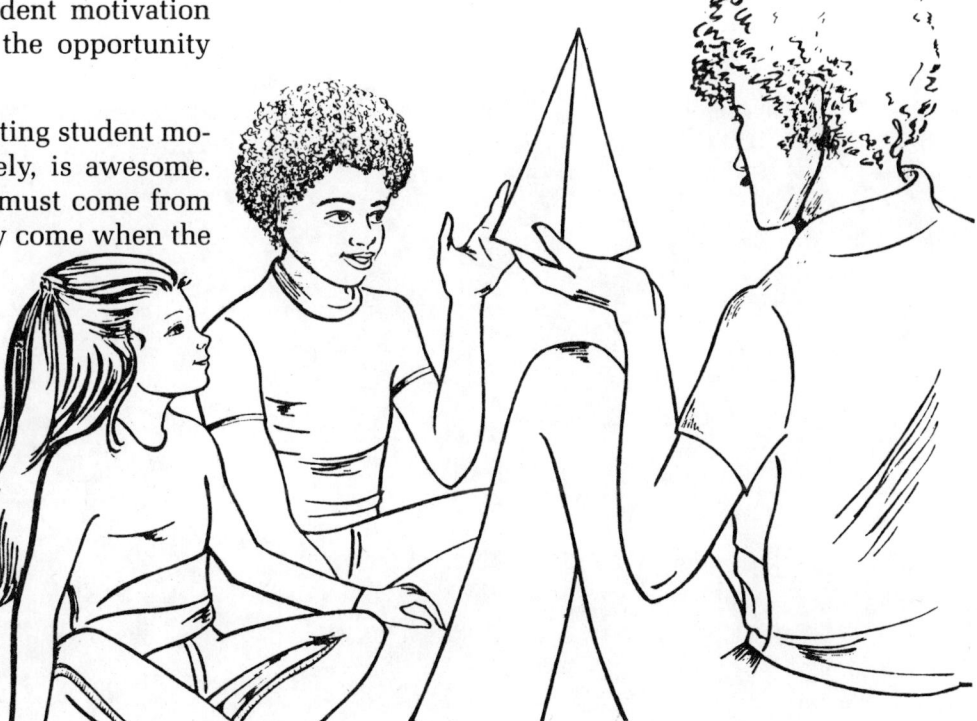

References (continued)

Perspectives for Change. Washington, D.C.: National Institute of Education, 1977.

* 7. Pearson, James R. "A Favorable Learning Environment." In *The Slow Learner in Mathematics*, 35th NCTM Yearbook. Reston, Virginia: National Council of Teachers of Mathematics, 1972.

8. Romberg, Thomas A. "The Diagnostic Process in Mathematics Instruction." In Jon Higgins and James Heddens (eds.), *Remedial Mathematics: Diagnostic and Prescriptive Approaches.* Columbus, Ohio: ERIC/SMEAC, 1976.

9. Suydam, Marilyn, and J. Fred Weaver. "Attitudes and Interests." In *Using Research: A Key to Elementary School Mathematics.* Columbus, Ohio: ERIC/SMEAC, 1975.

*The references with asterisks are readings we believe would be especially interesting to and useful for teachers.

Research Within Reach
Elementary School Mathematics

Meaning in Elementary School Mathematics

Question: A number of students in my sixth-grade class still have difficulty with techniques they have been exposed to for several years, like regrouping in subtraction. Can I help them by making my instruction more meaningful for their everyday lives? What are some ways to help students to retain the mathematics they learn?

In nearly fifty years of attention from educators, the concept of "teaching for meaning" has developed from the level of opinion to the level of acceptance—an acceptance firmly grounded in research. Through the lens of meaning, mathematics is seen as a closely knit system of understandable ideas, principles, and processes, and mathematics instruction is viewed as most effective when children see sense in what they learn. The challenge for teachers, in applying the theory to the classroom, is to determine what will be meaningful in particular instructional situations.

In the 1930's the meaning approach to mathematics teaching and learning became a rallying point for those who opposed the idea of teaching arithmetic through drill and memorization alone. In recent years, the focus has been on determining the appropriate balance between meaning and memorization in the learning and retention of mathematics.

Of course, the question which begins this bulletin gives the word "meaningful" a different cast. Rather than using it in the context of the *internal structure* of mathematics, the teacher refers to the *social significance* of mathematics learning in the lives of children.

In the early years of the theory, both interpretations of "meaningful" attracted the attention of educators (14). Through the decades, however, research studies and other investigations have provided evidence that a child is more likely to retain mathematical learning in an environment which highlights both *internal structure and meaning* and *applications and social significance* (6).

Retention of Basic Facts

In a meaningful development of basic facts knowledge, children move through a series of stages that begins with counting and ends with immediate recall (13). In the middle stages they combine counting with elements of guessing and solving from known combinations (for example, "3×5=15 and 5 more is 20, so 4×5=20"). For the mastery of basic facts, drill by itself has proven inferior to a combination of meaningful instruction and drill (14). The essential component of any such meaningful approach is the accommodation of children's need to pass through intermediate stages on the way to mastery of basic facts.

One meaningful approach to basic facts instruction was researched with second and fourth graders. The children were taught basic facts with a heavy emphasis on thinking strategies—the second graders for addition and subtraction and the fourth graders for multiplication and division. Examples for the two groups, respectively, follow:

> The mastery of doubles is followed by "doubles plus one" like 6+7 and 5+6. This is followed by "sharing numbers" like 7+9 and 5+7, which can be turned into doubles 8+8 and 6+6 by "sharing" the number 1.

> Mastery of multiplication by 2's, 5's, and 9's, using the patterns of those basic facts—for example, the sum of the digits in a basic 9-product is 9; the children build on those to get the other multiplication facts, exploiting patterns on the way, as in the following grid:

×	1	2	3	4	5	6
1						
2						
3						
4						
5						

The experiment revealed that the children readily adopted the strategies and continued to use them. A followup test compared their retention of basic facts with that of children taught in a traditional way; the experimental groups were clearly superior at both grade levels (15; also see 8 and 11).

Retention of Computational Skills

Teaching the computational algorithms for meaning cannot be a quick and easy enterprise. It has to be a measured sequence of steps that generates an increasing amount of meaning and culminates in the final step, the paper and pencil techniques that give answers to

$$\begin{array}{cccc} 135 & 104 & 65 & \\ +29 & -87 & \times 21 & 25\overline{)675} \end{array}$$

Generally, such a sequence should begin with the young child's introduction to the spirit of algorithmic manipulation on a very concrete level (see the articles by Merseth and Hazekamp in reference 12). As the concrete interpretations lead to pictorial and then symbolic representations, teachers can be working for meaning and retention in several ways:

- Teachers can work with calculators to complement work on paper and pencil algorithms. Used wisely, calculators can reinforce the awareness of meaning in mathematics. For example, with calculators that have the capability of repeated subtraction, teachers can give a stronger focus to meaning in division: "Enter 96 − 8 =. Keep pressing =. How many times must you press = to get 0?" (See reference 9 for further examples and check the *Research Within Reach* bulletin "Calculators in the Classroom.")

- Estimation and mental arithmetic must be used consistently as computational support: "You need to find out what 54 × 21 equals. Round both numbers to the nearest ten. How large is 50 × 20? About how large do you think 54 × 21 will be? What are your reasons?" (See the *Research Within Reach* bulletin "Estimation and Mental Arithmetic" for further suggestions.)

Role of Drill and Practice

One message that shines forth clearly from the research literature of the last few decades is the superiority of meaningful instruction over rote instruction for long-term retention of mathematical learning. In the short run, learning derived primarily from rote memorization might develop more efficiently, but that advantage diminishes with time (14). While drill is not a preferred way of *learning* mathematics, it can be an excellent way to *consolidate* mathematical learning—provided understanding comes first (1).

Several studies have looked at the effect of non-rote practice on the retention of mathematical learning. One of them, with possible implications at the elementary level, compared seventh-grade students on the retention effects of immediate versus delayed practice (five days after learning). The learning content came from rules for combining exponents (for example, $2^3 \times 2^4 = 2^7$); the practice consisted of working problems that demonstrated a previously learned rule.

From the results of a retention test administered three weeks after the original learning, the researcher concluded that delayed practice is more effective than immediate practice in enhancing long-term retention of mathematical rules (5).

Once again, we stress that this study was carried out at the seventh-grade level. For teachers of intermediate grades, there probably is some value in exposing students to the kind of practice that involves the retrieval of skills learned several days earlier, rather than practice that follows immediately on the heels of learning.

Another study called into question the traditional method of mathematics instruction whereby *all* students receive the *same amount* of practice. After creating an index of the optimal number of practice exercises for each student, the researchers divided the students into three groups: those for whom the index would determine the number of practice exercises during the experiment, those who would be free to choose the number of practice exercises, and those who would receive a fixed number of practice exercises throughout the experiment. The last group, the fixed practice group, compared unfavorably with the choice and index groups in a measure of long-term retention of mathematics (2). The message for teachers: take pains to individualize practice as much as possible.

Children are more likely to retain mathematical learning over time if meaningful, developmental activities comprise at least half of their mathematical instruction (and, it follows, drill and practice less than half). This intimate relationship among time, meaning, and retention has shown up in several investigations (one is detailed in reference 10).

Providing further evidence is a recent study that deserves attention because of its implications for structuring instruction in mathematics. The details of the study are important and are available in Good and Grouws' research report (3). In brief, the researchers studied effects on fourth-graders' mathematics achievement when a group of teachers followed a strict scheme of instruction that specified both teacher behaviors and time allotments: daily review (8 minutes); meaningful development of new material (30 minutes); seatwork (about 15 minutes); homework assignments; and special reviews (weekly, monthly). After two and one-half months, the students taught according to the experimental scheme achieved higher than comparable students in non-experimental classrooms. Clearly, teachers should aim for a *regular* and *consistent* blend of practice with meaningful instruction.

Besides those already noted, there are other suggestions, rooted in research and in the experiences of educators, for nurturing the relationship between meaningful instruction and retention.

Maintaining a Meaningful Environment

A good deal of research has been devoted to investigating discovery approaches to teaching mathematics. It appears that learning by discovery nourishes long-term retention of mathematics (17). However, teachers need not drastically change their teaching approach in order to gain from these investigations. Indeed, much of the benefit that accrues to children when they "discover" mathematical principles can also develop if teachers inject some discovery techniques into their normal instruction.

- Make children keen to recognize *patterns* whenever they arise (and manufacture situations in which they will arise).

- Make mathematical *generalizations* with the children whenever it is both possible and appropriate ("When you divide a decimal by ten, the decimal point gets moved one space to the left."), and teach them to do the same. Teachers need to beware of the children's developing false generalizations ("When you divide you always get a smaller number.").

- Point to the *connective tissue* that underlies all mathematics—that subtraction is the inverse of addition, that division is the inverse of multiplication, that multiplication is the addition of equal addends, and so on.

- Regularly ask *questions* like, "What could we do next?" "Is the answer reasonable?" "Could you make up a rule from these examples?" (For further suggestions on question-asking, see reference 6 and the *Research Within Reach* bulletin "Mathematical Problem Solving: Not Just a Matter of Words.")

- Alert the children to the appearances of mathematics in their everyday world. In pointing to the *social significance of* mathematics, teachers should not overemphasize raw computation. Problems with a consumer flavor, or problems with athletic, travel, or other interesting orientations can be made much more meaningful for children if they are presented in the language of estimation and approximation. (See reference 16 for further suggestions.)

Conclusion

Teaching for meaning in mathematics gives children an intellectual foundation to stand on and a familiar framework to build on. It also strengthens that framework by prolonging the retention of their mathematical learning.

References

* 1. Callahan, Leroy G., and Vincent J. Glennon. *Elementary School Mathematics: A Guide to Current Research.* Fourth edition. Washington, D.C.: Association for Supervision and Curriculum Development, 1975.

2. Gay, Lorraine R. "Use of a Retention Index for Mathematics Instruction." *Journal of Educational Psychology,* 63 (October 1972): 466-472.

* 3. Good, Thomas L., and Douglas A. Grouws. "The Missouri Mathematics Effectiveness Project: An Experimental Study in Fourth-Grade Classrooms." *Journal of Educational Psychology,* 71 (June 1979): 355-362.

* 4. Hazekamp, Donald W. "Teaching Multiplication and Division Algorithms." In Marilyn N. Suydam and Robert E. Reys (eds.), *Developing Computational Skills,* 1978 NCTM Yearbook. Reston, Virginia: National Council of Teachers of Mathematics (NCTM), 1978.

5. Horwitz, Stephen. "Effects of Amount of Immediate and of Delayed Practice on Retention of Mathematical Rules." Paper presented at the annual meeting of the American Educational Research Association, Washington, D.C.: April 1975. (ED 120 010)

6. Mathematics Resource Project. *Didactics and Mathematics.* Palo Alto, California: Creative Publications, Inc., 1978.

7. Merseth, Katherine Klippert. "Using Materials and Activities in Teaching Addition and Subtraction Algorithms." In M. Suydam and R. Reys (eds.), *Developing Computational Skills,* 1978 NCTM Yearbook. Reston, Virginia: NCTM, 1978.

8. Rathmell, Edward C. "Using Teaching Strategies to Teach the Basic Facts." In M. Suydam and R. Reys (eds.), *Developing Computational Skills,* 1978 NCTM Yearbook. Reston, Virginia: NCTM, 1978.

9. Reys, Robert E., et al. *Keystrokes: Calculator Activities for Young Students.* Palo Alto, California: Creative Publications, Inc., 1979.

References (continued)

10. Shuster, A.H., and F. Pigge. "Retention Efficiency of Meaningful Teaching." *The Arithmetic Teacher*, 12 (January 1965): 24-31.

11. Suydam, Marilyn N., and Donald J. Dessart. *Classroom Ideas from Research on Computational Skills.* Reston, Virginia: NCTM, 1976.

*12. Suydam, Marilyn N., and Robert E. Reys (eds.). *Developing Computational Skills.* 1978 NCTM Yearbook, Reston, Virginia: NCTM, 1978.

*13. Suydam, Marilyn N., and J. Fred Weaver. "Research on Mathematics Learning." In Joseph N. Payne (ed.), *Mathematics Learning in Early Childhood*, 37th NCTM Yearbook. Reston, Virginia: NCTM, 1975.

*14. Suydam, Marilyn N., and J. Fred Weaver. *Meaningful Instruction in Mathematics Education.* Columbus, Ohio: ERIC Clearinghouse for Science, Mathematics, and Environmental Education, 1972.

15. Thornton, Carol A. "Emphasizing Thinking Strategies in Basic Fact Instruction." *Journal for Research in Mathematics Education*, 9 (May 1978): 214-227.

16. Trafton, Paul R. "Estimation and Mental Arithmetic: Important Components of Computation." In M. Suydam and R. Reys (eds.), *Developing Computational Skills*, 1978 NCTM Yearbook. Reston, Virginia: NCTM, 1978.

17. Worthen, Blaine R. "A Study of Discovery and Expository Presentation: Implications for Teaching." In Robert B. Ashlock and Wayne L. Herman, Jr. (eds.), *Current Research in Elementary School Mathematics.* New York: The Macmillan Company, 1970.

*These are the readings we believe would be especially interesting to and useful for teachers.

Research Within Reach
Elementary School Mathematics

Securing Mathematical Skills: Drill and Other Topics

Question: How do you determine an appropriate blend of manipulatives and drill in teaching? When should drill be used to reinforce the understanding students get from their work with manipulatives?

Though different from adults in many aspects of learning, children share with us one overriding characteristic. We all learn most efficiently when what we learn interests us, seems useful to us, and is presented to us, not in random fashion, but organized in a pattern or order that we can grasp.

In a subject like mathematics, where concepts and skills are learned in sequence and where the quality of learning often depends on quick and regular recall, teachers face a special challenge. In addition to knowing how to organize material for efficient learning and how to portray it as both interesting and useful, mathematics teachers must know when to press for instant recall and when *not* to, when to review and how much to review, and what kind of balance to strike between teaching new information and practicing acquired skills.

The *Research Within Reach* bulletin "Meaning in Elementary School Mathematics" concentrates on ways to help students see sense and meaning in mathematics. We now look at the next step—what research and experience have to say about *reinforcing* learning in elementary school mathematics. In particular, we will look at drill, at homework and review, and at the variety of in-class activities commonly termed "practice."

Drill

Drill is the teaching strategy that uses repetition to develop precision in learning and to fix facts for efficient recall. Both research and experience have confirmed the value of well-designed and appropriate drill in elementary school mathematics. Understanding the particulars of "well-designed and appropriate," however, requires at least a brief glance at the last half-century of teaching and research in elementary school mathematics.

Fifty years ago there was considerable support for *teaching* by drill. Elementary school mathematics was ideally suited to a stimulus-response approach to learning. Whether concepts were understood or skills had meaning was of secondary concern to the proponents of this position.

Another group of educators, led by W. A. Brownell, was opposed to the learning-by-drill position. They maintained that children need to see sense and meaning in what is learned before drill can be very effective. Their research, which has continued through the decades, has substantiated their position and has helped to define the role of drill in mathematics education—a role in which *drill complements teaching for meaning and is carefully balanced with other forms of practice and with the regular learning of new material*.

Before Drill Begins

The overwhelming consensus among research findings is that drill in basic facts should begin only after relevant concepts, like number, addition, or subtraction, have been developed in a meaningful way for children. For example, children should see some relation between $5 + 6 = 11$ and $4 + 7 = 11$ and not just commit those facts to memory.

Given that children should not begin to memorize basic facts until they have some understanding, what should teachers look for as *signs* that a particular child is ready for memorization? Davis (8) has provided some clear indicators. Using $5 + 4 = 9$ as a sample, he stresses that the teacher should watch, first of all, for signs that the child understands *what* the fact says:

1. The child can create or recognize embodiments of the fact (she can use the number line, rods, counters, or fingers to show $5 + 4 = 9$).
2. The child can understand the concepts in the fact (he characterizes and describes the symbols denoted by 5, 4, 9, +, and =).
3. The child can use the fact in simple exercises (for example, $5 + 4 + 1 = ?$).
4. The child can make up a story problem using $5 + 4 = 9$.

At the next level are signs that the child understands *why* the fact is true:

5. The child can show the truth of the fact using objects, models, or other facts (for example, starting with $4 + 4 = 8$, can she show that $5 + 4 = 9$? or can she separate a set of 9 objects into a set of 5 objects and a set of 4 objects?).
6. The child can complete related statements of the number fact ("What makes $5 + \square = 9$ true?").

If a child shows *most* of these six signs, then it is time for memorization, and Davis (8) provides some principles for memorization, summarized here:

- Have children begin to memorize basic arithmetic facts *soon* after they demonstrate an understanding of symbolic statements.

- Children should participate in drill with the intent to memorize.

- During drill sessions, emphasize remembering —don't explain!

- Keep drill sessions short and have some drill almost every day.

- Try to memorize only a few facts in a given lesson and constantly review previously memorized facts.

- Express confidence in your students' ability to memorize—encourage them to try memorizing and see how fast they can be.

- Emphasize verbal drill exercises and provide feedback immediately.

- Vary drill activities and be enthusiastic.

- Praise students for good efforts—keep a record of their progress.

There is a strong research base for all of these principles, and a few merit some elaboration. For instance, in order for students to begin drill with the intention of memorizing, it is helpful for teachers to discuss the goals of drill with them (25). (Examples of goals: the students are taking part in drill so that they will be able to attack longer computations with a minimum of energy directed to the underlying basic facts; or, they are drilling so that they will be able to make rough but reasonable estimates without the distractions involved in computing basic facts.)

There is some evidence that, once a child can represent basic facts concretely—with blocks, rods, or other manipulatives—teachers should lay the groundwork for Davis's drill principles in two ways:

- First, teachers should take into account *each child's* own particular counting strategies. When children begin to count they do it in rote fashion, with no reference to real objects. Eventually, their counting becomes quite sophisticated ("24 + 5 = ? Let's see, 24 . . . 25, 26, 27, 28, 29. So, 24 + 5 = 29."). In between rote counting and such counting-on strategies there are several stages of counting, and the development of counting strategies can vary from child to child. What is predictable, however, is the influence counting strategies have on a child's efforts to learn basic facts, and teachers can and *should* use these strategies as they help each child master the basic facts. For more information on counting, see the *Research Within Reach* bulletin "Counting Strategies" and references 11, 16, 20, and 24.

- Second, teachers should organize and present basic facts in a way that makes them easier to remember. There are several sources of guidance for basic facts organization (8, 13). Folsom (9) recommends that first grade teachers group the 0+ and +0 facts together to stress the point that adding zero to a number, or adding a number to zero, doesn't change the number. Next come the 1+ and +1 facts, then the doubles (4 + 4 = 8). Finally, teachers must point out, on a *regular* basis, that it doesn't matter what order numbers are listed in when they are added—in other words, the property of commutativity. Teachers must do the same for multiplication, when that is the focus of attention.

- Later, probably in the second grade, teachers can work to get children to recognize the doubles-plus-one facts (for instance, "4 + 5 is 4 + 4 + 1, and that is 8 + 1, or 9"), and then the regularity in the 9-facts (9 + 1 ends in 0, 9 + 2 ends in 1, 9 + 3 ends in 2).

How strongly teachers should stress such thinking strategies is still an area of active research (7, 22), but there is enough evidence of beneficial effect, especially in the area of retention (24), to warrant regular attention to the teaching of thinking strategies.

When and How to Drill

There are a number of sources of interesting drill ideas, using flash cards, games, calculators, peer tutoring, etc. (See, for example, references 3, 8, 9, 21, and 25, as well as regular issues of the *Arithmetic Teacher.*)

As you attend to Davis's drill principles listed above, keep in mind the following related research findings:

- Daily, brief drill, in *conjunction with regular review and the meaningful development of new material*, is effective in improving mathematics achievement (12).

- Along with its many other classroom roles, the handheld calculator has emerged as an exciting tool for immediate feedback in basic facts drill work (3, 6, 18, 23).

- Consistent teacher praise is an important ally in effective drill work (23).

- The effectiveness of devices like flash cards is heightened if they are used when children are grouped according to basic facts needs. Regular timed tests can reveal what those groupings should be (9).

- It is important to confine drill to appropriate topics and times. A memorization approach to problem solving, for example, is definitely counterproductive (5), and research has implied that routine drill homework may not be of much value (2).

Other Ways to Reinforce Learning

Even though students work most efficiently with computational algorithms when they know the basic facts at a stimulus-response level, teachers need not wait for complete basic facts mastery in order to begin meaningful instruction in the use of algorithms. In fact, it is appropriate to practice basic facts *during*, as well as *after*, the learning of computational skills. Ashlock and Washbon (1) provide a number of basic facts game activities for just this kind of setting.

What kind of scheduling leads to good practice? A recent study (12) considered the effects on fourth-grade mathematics achievement when a group of teachers followed a strict scheme of instruction that specified both teacher behaviors and time allotments for daily review (8 minutes); meaningful development of new material (30 minutes); seatwork involving uninterrupted, successful practice that is checked at the end of the period (about 15 minutes); homework assignments that involve about 15 minutes of work at home and include several review problems; and special reviews (weekly, monthly). After two and one-half months, the students taught according to the experimental scheme achieved higher than comparable students in non-experimental classrooms. Clearly, teachers should aim for a *regular* and *consistent* blend of meaningful instruction with review, seatwork, and homework.

Seatwork should *not* consist only of paper-and-pencil practice. Teachers should expect and encourage students to make an estimate of the answer *before* they approach each problem (18):

Start with   28
                           +31

Then,   28   "28 is about _____"
               +31   "31 is about _____"

Finally,   28
               +31

"is about _____" (the sum of the estimates)

After their computations, students should regularly answer the question, "Does my answer make sense?" Teachers should help students to recognize what they must do if they get an answer that does not make sense.

As pointed out in the *Research Within Reach* bulletin "Estimation and Mental Arithmetic," a child's mathematical development can be made richer and deeper through regular practice in mental arithmetic skills, such as rounding, adding from left to right, and subtraction with the help of compensation (for example, $142 - 96$ can be figured out mentally by adding a 4 to each term to give $146 - 100$, or by doing $142 - 100$, then adding 4).

Calculators can and should play a *central* role in classroom practice, and computation is only one area where they can contribute to the growth of learning. Once a student understands the workings of a particular algorithm, a calculator can help cement that understanding. For example,

Ask the children to divide 649 by 4 with their calculators and to determine from the machine's answer (a whole number and a decimal) what the remainder would be in a paper-and-pencil computation.

Even among older students, the inability to handle such an exercise is widespread. A similar problem appeared on the most recent National Assessment of Educational Progress (NAEP) in mathematics:

> *A man has 1310 baseballs to pack in boxes which hold 24 baseballs each. How many baseballs will be left over after the man has filled as many boxes as he can?*

Of the thirteen year olds who worked the problem without calculators, only 29 percent did it correctly. In the groups who used calculators to solve the problem, however, the results were *even worse:* only 6 percent of the thirteen year olds and 19 percent of the seventeen year olds came up with the correct answer (19). Clearly, our schools must provide earlier and more regular practice in using calculators to deepen conceptual understanding.

For further information on the use of calculators, see Wheatley, et al. (26), and the *Research Within Reach* bulletin "Calculators in the Classroom."

The effectiveness of mathematics homework was the focus of Austin's comprehensive survey of the research literature (2). Although no homework studies at the second-grade level or below are reported, this research concluded that homework seems to improve computational skills. Austin also found that comments on homework can improve student achievement, and that it is not necessary to grade *every* homework exercise.

As in the case of drill, daily assignments that are short (around 15 minutes) are the best use of practice, and long-term retention is best served if assignments around a particular skill are spread out in time, rather than concentrated within a short interval of time (17). Thus, it would be better to review a skill or concept one day and seven days after the initial learning, rather than scheduling the reviews in the first two days after the learning. Furthermore, research suggests that practice of mathematical rules that follows immediately on the heels of learning is not as effective as practice that is delayed for several days (15).

Research also confirms that the amount of practice should vary from student to student. One study (10) compared the long-term retention of mathematical skills among three groups of students—students who were free to choose their amounts of practice, students whose optimal amounts of practice had been individually determined beforehand, and students who received a fixed amount of practice. The fixed-practice group compared unfavorably with both the choice group and the individualized-practice group in a measure of long-term retention. The message for teachers: *take pains to individualize practice as much as possible.* Some suggestions for doing this can be found in the *Research Within Reach* bulletins "Diagnosis: Taking the Mathematical Pulse" and "Learning Elementary School Mathematics: Individual Styles and Individual Needs."

One key factor in the design of individualized practice is *success*—taking care that the activities a child tackles have the potential for a high degree of success. One extensive study of classrooms at the second- and fifth-grade levels (4) found that students who spend over half of their time working on high-success tasks have higher achievement scores in both reading and mathematics. The findings indicated that the need for high-success practice is especially great in the lower grades and that the amount of time needed for such practice decreases at the fifth-grade level and beyond.

Success is important, but that does *not* mean a teacher should assign "more of the same" after a child has clearly mastered a particular skill. It is always possible to use the skill being practiced as a springboard to challenging problem-solving experiences (18). For example, if the practice has been aimed at the multiplication algorithm, and the student has moved smoothly through her practice exercises, then the teacher, rather than asking her to solve 63 × 12 = ?, might ask her to fill in the blanks in the following problem with the appropriate digits:

$$\begin{array}{r} 6\,? \\ \times\,1\,2 \\ \hline 1\,2\,? \\ 6\,? \\ \hline 7\,5\,? \end{array}$$

Conclusion

To be most effective, mathematics teachers must apply a measured blend of meaning and reinforcement to each topic they present. The doses of reinforcement, in the form of drill, review, homework, and non-rote classroom practice, should be well-timed, brief, and regular. Used carefully by teachers, they need not become the tedious rituals which too many children perceive them to be. Instead, they can secure mathematical skills already learned and open the way to further learning.

References

* 1. Ashlock, Robert B., and Carolynn A. Washbon. "Games: Practice Activities for the Basic Facts." In Marilyn N. Suydam and Robert E. Reys (eds.), *Developing Computational Skills*, 1978 NCTM Yearbook. Reston, Virginia: National Council of Teachers of Mathematics (NCTM), 1978.

2. Austin, Joe Dan. "Homework Research in Mathematics." *School Science and Mathematics*, 79 (February 1979): 115-121.

* 3. Beardslee, Edward C. "Teaching Computational Skills with a Calculator." In M.N. Suydam and R.E. Reys (eds.), *Developing Computational Skills*, 1978 NCTM Yearbook. Reston, Virginia: NCTM, 1978.

4. Beginning Teacher Evaluation Study. *Final Report*. San Francisco: Far West Laboratory for Educational Research and Development, 1978.

* 5. Callahan, Leroy G., and Vincent J. Glennon. *Elementary School Mathematics: A Guide to Current Research*. 4th edition. Washington, D.C.: Association for Supervision and Curriculum Development, 1975.

6. Channel, Dwayne E. *The Use of Hand Calculators in the Learning of Basic Multiplication Facts*. Columbus, Ohio: The Calculator Information Center, 1978.

7. Cifarelli, Victor V., and Grayson H. Wheatley. "Formal Thinking Strategies: A Prerequisite for Learning Basic Facts?" *Journal for Research in Mathematics Education*, 10 (November 1979): 368-370.

* 8. Davis, Edward J. "Suggestions for Teaching the Basic Facts of Arithmetic." In M.N. Suydam and R.E. Reys (eds.), *Developing Computational Skills*, 1978 NCTM Yearbook. Reston, Virginia: NCTM, 1978.

9. Folsom, Mary. "Operations on Whole Numbers." In Joseph N. Payne (ed.), *Mathematics Learning in Early Childhood*, 37th NCTM Yearbook. Reston, Virginia: NCTM, 1975.

10. Gay, Lorraine R. "Use of a Retention Index for Mathematics Instruction." *Journal of Educational Psychology*, 63 (October 1972): 466-472.

*11. Gibb, E. Glenadine, and Alberta M. Castaneda. "Experiences for Young Children." In J.N. Payne (ed.), *Mathematics Learning in Early Childhood*, 37th NCTM Yearbook. Reston, Virginia: NCTM, 1975.

12. Good, Thomas L., and Douglas A. Grouws. "The Missouri Mathematics Effectiveness Project: An Experimental Study in Fourth-Grade Classrooms." *Journal of Educational Psychology*, 71 (June 1979): 355-362.

*13. Hazekamp, Donald W. "Teaching Multiplication and Division Algorithms." In M.N. Suydam and R.E. Reys (eds.), *Developing Computational Skills*, 1978 NCTM Yearbook. Reston, Virginia: NCTM, 1978.

14. Heddens, James W. "A Theoretical Study of the Organization of Basic Addition Facts for Memorization." In Mary Ellen Hynes (ed.), *Topics Related to Diagnosis in Mathematics for Classroom Teachers*. Bowling Green, Ohio: Research Council for Diagnostic and Prescriptive Mathematics, 1979.

15. Horwitz, Stephen. *Effects of Amount of Immediate and of Delayed Practice on Retention of Mathematical Rules*. Paper presented at the annual meeting of the American Educational Research Association, Washington, D.C., April 1975, ED 120 010.

16. Leutzinger, Larry P., and Glenn Nelson. "Let's Do It: Counting With A Purpose." *Arithmetic Teacher*, 27 (October 1979): 6-9.

*17. Mathematics Resource Project. "The Teaching of Skills." In *Didactics and Mathematics*. Palo Alto: Creative Publications, Inc., 1978.

18. McKillip, William D., and Cherie Adler Aviv. "How to Use, Not Abuse, Those Practice Exer-

cises." *Arithmetic Teacher*, 26 (April 1979): 10-13.

19. National Assessment of Educational Progress. *The Second Assessment of Mathematics, 1977-78: Released Exercise Set*. Denver, Colorado: Education Commission of the States, 1979.

20. Rathmell, Edward C. "Using Teaching Strategies to Teach the Basic Facts." In M.N. Suydam and R.E. Reys (eds.), *Developing Computational Skills*, 1978 NCTM Yearbook. Reston, Virginia: NCTM, 1978.

21. Smith, Seaton E., Jr., and Carl A. Backman (eds.) *Games and Puzzles for Elementary and Middle School Mathematics: Readings from the Arithmetic Teacher*. Reston, Virginia: NCTM, 1975.

22. Steffe, Leslie P. "A Reply to 'Formal Thinking Strategies: A Prerequisite for Learning Basic Facts?'" *Journal for Research in Mathematics Education*, 10 (November 1979): 370-373.

*23. Suydam, Marilyn N., and Donald J. Dessart. *Classroom Ideas from Research on Computational Skills*. Reston, Virginia: NCTM, 1976.

24. Thornton, Carol A. "Emphasizing Thinking Strategies in Basic Fact Instruction." *Journal for Research in Mathematics Education*, 9 (May 1978): 214-227.

25. Trafton, Paul R., and Marilyn N. Suydam. "Computational Skills: A Point of View." *Arithmetic Teacher*, 22 (November 1975): 528-537.

26. Wheatley, Grayson H., et al. "Calculators in Elementary Schools." *Arithmetic Teacher*, 27 (September 1979): 18-21.

*The references with asterisks are readings we believe would be especially interesting to and useful for teachers.

Research
Within
Reach
Elementary School Mathematics

Grouping for Elementary School Mathematics

Question: I would like to do some grouping in my class but am a bit confused. Would it be productive for me to keep my students in the groups that form naturally in September?

Are there guidelines for organizing the classroom so that every child has an optimal chance to learn mathematics? The answer is yes, but because every child approaches mathematics with a unique set of learning needs, the guidelines proposed are neither rigid nor universal. Instead, they stress teacher flexibility as a hallmark of effective classroom organizing and fix the roots of that flexibility in an awareness of children's various needs and of the options for meeting those needs.

A recent national survey, however, indicates that many teachers ignore the value of flexibility. The survey's data suggest "very common use of an instructional style in which teacher explanation and questioning is followed by student seatwork on paper and pencil assignments" (9, p. 12). Among the trends that appeared is "a steady decrease in recent years in the use of manipulatives and student involvement techniques" (9, p. 12).

Teachers who minimize the classroom involvement of small children are locked into a strategy that is bound to fail many of those children. They ignore proven alternatives, like grouping, that enhance opportunity and allow children to develop naturally in mathematics.

Grouping strategies have long been a subject of research. Some studies have looked at whole classes of students grouped by general ability (I.Q. or overall grade-point average), while others have considered whole-class groupings by previous mathematical achievement. In some studies, students were grouped homogeneously (that is, they were similar in ability or achievement), while in others the groupings were heterogeneous. These whole-class studies have not yielded conclusive evidence (3, 17). From his survey of the research literature, Begle concluded that if students are to be grouped homogeneously, then mathematics achievement, rather than general ability, should be the criterion. He pointed out that homogeneous grouping is of questionable value for students in the low to middle range of mathematics achievement. High achievers, on the other hand, tend to show an increase in mathematics achievement as a result of homogeneous grouping (3).

Some educators argue against fixed groupings by ability or achievement on the grounds that the strategy could have destructive effects on student attitude and self-concept. On a broad scale no such effects have shown up in the research literature (3). However, a recent study (5) has raised the possibility that the negative effects work on a more subtle level. In that study, seventh graders were placed in three groups (high, medium, and low) according to scores on the California Test of Basic Skills. After a trimester with these fixed groups the researchers looked closely at several aspects of student attitude. Among their conclusions: low-ranked students in *each* ability level appeared to have the highest anxiety toward mathematics; the low-ranked students in the middle-level groups had the highest anxiety of all, the lowest self-concept, and the least enjoyment of mathematics.

Rather than focusing attention on fixed ability or achievement groups as an overall approach to classroom organization, Begle recommends that teachers experiment with intra-class grouping, in particular, with regrouping at the beginning of each new topic (3). This is no small challenge, and each teacher will have to make adjustments based upon experimentation, but the strategy holds the promise of effective attention to individual differences in the learning of mathematics.

Bierden (4) outlines three requirements for successful intra-class grouping: (1) flexibility; (2) accuracy; and (3) appropriate materials. Together, they provide a convenient scheme for presenting what research and practice have to say about intra-class grouping.

Flexibility

To be successful with intra-class grouping, teachers should have *short-term goals* for their various small groups, they should *discuss those goals* with the students in each group, and they should *regroup frequently* as the goals are met and fresh goals present themselves. The range of appropriate goals is limitless. Learning basic facts, practicing problem-solving techniques, reviewing recent learning, doing remedial work —all are appropriate for small-group work.

Basic facts offer an excellent opportunity for flexible grouping, and they even offer a convenient technique for the grouping process—the teacher tests the students regularly on basic facts and then groups them according to the basic facts they do not know (10). Thus, one group might contain five children who share a weakness in the multiplication facts involving 7 and 8.

Small, flexible groups make an ideal setting for problem-solving work. As pointed out in the *Research Within Reach* bulletin "Mathematical Problem Solving: Not Just a Matter of Words," successful problem solvers have several problem-solving skills in common. Among them are the ability and willingness to produce a variety of solution avenues, a facility for changing solution avenues when dead-ends appear, and the ability to look back at a solution process for mistakes and inconsistencies. As children interact with one another in small groups, these skills develop from their interactions, with a little help from the teacher. For example, if a problem-solving competition is set up among small groups, and penalty points are assigned when a group says it is sure of an answer that turns out to be incorrect, then there is an incentive for looking back and checking. The teacher then has the opportunity to show each group how to cooperate in making sure the checking is thorough. (See also reference 13.)

Reviewing is another area that invites small-group work as a supplement to whole-class review. Ashlock and Washbon suggest a number of games for practicing basic facts that can be used with large groups or small (1). Dunn and Dunn (8) describe the use of a "circle of knowledge" in which small groups of students cooperate in reviewing. For example, a group of five children might review addition by working together, each taking successive turns, to answer the question, "How many examples can you create where two numbers added together make 100?"

Grouping for remedial work (see reference 12) is especially dependent on a careful assessment by the teacher of the needs of individual students, and that leads to the second requirement, accuracy.

Accuracy

Whatever the learning goal for a particular small group, it is important that the students be engaged in work that has a good chance of being *successful* for them. The Beginning Teacher Evaluation Study (BTES), undertaken by the Far West Laboratory for Educational Research and Development, has recently underscored this in its extensive observations of teacher effectiveness at the second- and fifth-grade levels (2). The BTES researchers decided that a high success rate is important throughout elementary school, but especially at the primary level. They concluded that primary-level children achieve at a higher rate if approximately 50 to 75 percent of their time is spent on high-success tasks. The researchers also concluded that students who spend more time in small group instruction *with frequent teacher interaction* have higher rates of engagement on tasks than students who spend most of their time on independent seatwork and have little interaction with their teacher (2, 15).

It is important for teachers to have the kind of regular contact with students that allows them to decide what each child's current mathematical needs are, and where the child can experience a high degree of success in meeting those needs. To this end, the small flexible group offers a double advantage. Not only can teachers group students with an eye toward maintaining high-success tasks, but, in moving from group to group, the teacher has the opportunity to give each student more concentrated attention than is normally possible in a whole-class setting. For further suggestions on diagnosing student skills and weaknesses, see the *Research Within Reach* bulletin "Diagnosis: Taking the Mathematical Pulse."

Appropriate Materials

A picture of children's needs, no matter how accurate, is incomplete if the materials they work with do not offer a reasonable chance for success. Of particular concern is cognitive development, the progression each child makes from understanding concepts only on a concrete level to understanding at an abstract level. In one classroom, for example, there may be children who can tell you that $5 \times 16 = 80$ because 5 tens are 50 and 5 sixes are 30, so $5 \times 16 = 50 + 30 = 80$. There may also be children in the same class who cannot respond in that way, but who can explain $5 \times 16 = 80$ by counting from a pictorial representation of blocks. Finally, there may be children who need to work with real blocks, or similar items, in order to understand why $5 \times 16 = 80$. If an exercise like $5 \times 16 = ?$ is given to small groups, then it is essential that each group have a variety of materials on hand so that each child can settle on an appropriate level of meaning. As a matter of fact, teachers who are aware of developmental levels in their classrooms can group children according to those levels. In this way, the children who are grouped by developmental level can begin to understand concepts by watching each other, as well as by doing individual work.

Teachers who would like to look more closely at options for appropriate instructional materials can check references 6, 8, and 16 and the *Research Within Reach* bulletins "The Bridge from Concrete to Abstract" and "Meaning in Elementary School Mathematics."

Whole-Class Instruction

Just as teachers need to be flexible about reassigning students from one small group to another, so should they be flexible about moving from whole-class instruction to small-group instruction. There are occasions when it is appropriate for a teacher to be working with an entire class—for instance, when new materials or new topics are being introduced, or when the work of small groups on similar topics can be integrated. (Suppose groups A, B, and C have been working, separately, to measure the volumes of three different boxes. The teacher calls them together, has them record their answers on the board, and says, "Group A's box is 10 inches long, 8 inches high, and 4 inches wide. Can you estimate the dimensions of the other two boxes by using the recorded volumes?")

Rising and his colleagues (14) have argued that communication is at the heart of effective mathematics teaching—including questions from the students, questions to the students, discussions among the students—and that a whole-class setting is often very conducive to this kind of communication. (It is essential to distinguish, as do Rising and his colleagues, between "whole-class setting" or "teacher-centered classroom," on the one hand, and "lecture approach," on the other. No responsible educator defends lecturing as a teaching approach to small children.) In a recent study, Good and Grouws have looked at communication in a whole-class setting (11). They trained half of a group of fourth-grade teachers to approach their mathematics classes with a daily regimen of review, development of new concepts, and seatwork. The other teachers taught as usual. The regimen specified not only the amount of time spent on each of the tasks but also the kinds of teacher-student interactions that should take place at each point during the class. Overall, the teachers who used the daily regimen, all of whom taught mathematics to whole classes, brought a greater gain in achievement to their classes than did the other teachers, including those teachers who taught according to a small-group model.

Conclusion

The research literature offers no final solutions on the issue of organization for mathematics instruction. There are no final solutions. It may not be simple, but teachers must experiment with options, such as small groups, until they arrive at the organizational patterns that fit their goals and their particular students.

The implications of this review of research are varied, but one thing does seem clear. No rigid model of classroom organization—group, whole-class, or otherwise—can guarantee the best setting for learning mathematics. In the end, it is the teacher, with the freedom to match teaching strategies with children's needs, that determines the quality of learning in the classroom.

References

* 1. Ashlock, Robert B., and Carolynn A. Washbon. "Games: Practice Activities for the Basic Facts." In Marilyn N. Suydam and Robert E. Reys (eds.), *Developing Computational Skills*, 1978 NCTM Yearbook. Reston, Virginia: National Council of Teachers of Mathematics (NCTM), 1978.

2. Beginning Teacher Evaluation Study (BTES). *Final Report*. San Francisco, California: Far West Laboratory for Educational Research and Development, 1978.

3. Begle, E.G. *Ability Grouping for Mathematics Instruction. A Review of the Empirical Literature. SMEGS Working Paper No. 17.* 1975. [ED 116 938]

4. Bierden, James E. "Behavioral Objectives and Flexible Grouping in Seventh-Grade Mathematics." *Journal for Research in Mathematics Education*, 1 (November 1970): 207-217.

5. Brassell, Anne, et al. "Ability Grouping, Mathematics Achievement, and Pupil Attitudes toward Mathematics." *Journal for Research in Mathematics Education*, 11 (January 1980): 22-28.

References (continued)

* 6. Cathcart, W. George (ed.). *The Mathematics Laboratory: Readings from the Arithmetic Teacher.* Reston, Virginia: NCTM, 1977.

* 7. Crosswhite, F. Joe, and Robert E. Reys (eds.). *Organizing for Mathematics Instruction.* 1977 NCTM Yearbook. Reston, Virginia: NCTM, 1977.

8. Dunn, R.D., and K. Dunn. *Teaching Students through Their Individual Learning Styles: A Practical Approach.* Reston, Virginia: Reston Publishing Company, Inc., 1978.

9. Fey, James T. "Mathematics Teaching Today: Perspectives from Three National Surveys." *Arithmetic Teacher,* 27 (October 1979): 10-14.

10. Folsom, Mary. "Operations on Whole Numbers." In J.N. Payne (ed.), *Mathematics Learning in Early Childhood,* 37th NCTM Yearbook. Reston, Virginia: NCTM, 1975.

*11. Good, Thomas L., and Douglas A. Grouws. "The Missouri Mathematics Effectiveness Project: An Experimental Study in Fourth-Grade Classrooms." *Journal of Educational Psychology,* 71 (June 1979): 355-362.

12. Junge, Charlotte W. "Adjustment of Instruction (Elementary School)." In *The Slow Learner in Mathematics,* 35th NCTM Yearbook. Reston, Virginia: NCTM, 1972.

*13. Krulik, S. (ed.). *Problem Solving in School Mathematics,* 1980 NCTM Yearbook. Reston, Virginia: NCTM, in press.

14. Rising, Gerald R.; Stephen I. Brown; and Lawrence N. Meyerson. "Teacher-Centered Classroom." In F. Joe Crosswhite and Robert E. Reys (eds.), *Organizing for Mathematics Instruction,* 1977 NCTM Yearbook. Reston, Virginia: NCTM, 1977.

15. Schneider, E. Joseph. "Researchers Discover Formula for Success in Student Learning." *Educational R&D Report,* 2 (Fall 1979): 1-6.

*16. Suydam, Marilyn N., and Robert E. Reys (eds.) *Developing Computational Skills,* 1978 NCTM Yearbook. Reston, Virginia: NCTM, 1978.

*17. Suydam, Marilyn N., and J. Fred Weaver. *Using Research: A Key to Elementary School Mathematics.* Columbus, Ohio: ERIC Clearinghouse for Science, Mathematics and Environmental Education, 1975. Also available from NCTM.

*These are the readings we believe would be especially interesting to and useful for teachers.

Research Within Reach
Elementary School Mathematics

Learning Elementary School Mathematics: Individual Styles and Individual Needs

Question: In my class there are thirty children. I want to do as much as possible to attend to individual needs. Is there an effective way to individualize instruction in such a large class?

No rigid format for teaching mathematics can satisfy everyone's needs. When we learn, each of us favors individual inclinations and individual styles. For example, a quick memory and impulsive behavior can spell success for children whose teacher relies heavily on drill; similarly, if the teaching approach is strictly individualized or self-paced, the advantage often falls to the children who thrive on working alone.

Research has answered some questions and raised many others concerning the role of individual style in the learning of mathematics. One conclusion is clear, however: *educators must avoid the trap of assuming that a child's lack of success is due to a lack of talent. It could very well be due to a lack of opportunity or to an improper match between instruction and student learning style.*

Self-Paced Instruction

In using the word "individual" we are not tying it to the individualized, or self-paced, programs that are used in some schools. Such programs account for only one approach to accommodating individual differences among children. In those programs, students participate in setting individual goals, work at individual rates (either alone or in small groups), and have a part in evaluating their progress.

Critics claim that the only individualized feature within individualized programs is pace, since the same materials are covered in the same manner—through worksheets or programmed texts (14). The National Advisory Committee on Mathematical Education (NACOME) has reported other common objections to self-paced instruction: that the program emphases have remained in the area of computation; that there is little, if any, peer interaction; and that the stress on testing until mastery leads students to emphasize lower level skills, with little attention to higher analytic and problem solving skills (14).

Schoen reviewed many reports of individualized mathematics programs. He concluded that the extra expense of self-pacing and the extra time and effort required of teachers were not balanced by the effectiveness of the programs. In general, self-paced programs fared no better than traditional programs in mathematics achievement, and in the higher elementary grades there were signs that self-paced programs fared *worse* than traditional programs (17).

Other doubts about self-pacing focus on peer interaction. There is a good deal of evidence—from the research on child development of Piaget and others and from recent classroom observation studies—that the quality and quantity of peer interactions play a vital role in *every* child's school experience. In Piaget's portrayal of cognitive development, children need to interact with peers in order to hear the views of others and to clarify their own. Furthermore, according to Piagetian research, it is through the exchange of viewpoints that children develop their language and their ability to think logically (12, 13).

Besides affecting cognitive development, peer interactions can affect classroom performance and achievement. That is one of the conclusions of the extensive second- and fifth-grade observation studies of the Beginning Teacher Evaluation Study (BTES) undertaken by the Far West Laboratory for Educational Research and Development. These studies have shown that a child's level of achievement, including mathematics, is correlated with the amount of time engaged in learning tasks that have a high potential for success. This is especially true in the primary grades. The researchers further concluded that the amount of this engaged time is also related to the amount of time spent with peers and teachers in interactions about academic content. As one summary of the research has put it, "Students who work together to reach academic goals and take responsibility for achieving them generally have higher achievement" (16, p. 4).

The BTES results have a much wider scope than is portrayed here, and interested readers are encouraged to read the BTES literature (1). Other sources that address the need for *all* elementary school learners to interact while they learn mathematics are the *Research Within Reach* bulletins, "Grouping for Elementary School Mathematics," "Motivation in Mathematics," and "Mathematical Problem Solving: Not Just a Matter of Words."

Individual Learning Styles

Individual attitudes, preferences for content, special brain processes each person uses to absorb information—all of these we group under the umbrella term "individual learning styles." Psychologists have studied learning styles much longer than educators have, and that lag in time is reflected in the current state of the art in mathematics education research: more is known about what learning styles are than what teachers can or should do about them.

If research hasn't drawn any tight connections between learning styles and instruction, why should teachers bother about learning styles? There are no guarantees that an awareness of style will improve mathematics instruction, but there are some exciting possibilities. Standing to benefit, for example, is the child whose major impediment to learning mathematics is her style of learning, not an *inability* to learn. Once this distinction is made by the teacher, a mathematics program may take on a whole new character for that child, one in which success is more likely.

Sharp observation and diagnosis and the willingness to provide students with a variety of learning options are the keys to capitalizing on what is known about learning styles. Students working with mathematics might reveal some of the following differences in learning style:

- Some children are primarily tactile in their processing of information, others primarily aural, and others primarily visual. These *sense preferences* have a bearing on the learning of mathematics (7, 10). Similarly, children vary in their abilities to find meaning from various *symbols*: written words and spoken words; written numerical symbols and spoken numerical symbols (19). Behavior is the key indicator. Does a student appear to learn by reading? By listening? Are written problems solved more easily than spoken problems?

- Some children become *anxious* if they don't receive regular guidance from teachers or peers, while others prefer to work *independently* (21).

- Some children show what psychologists call field independence and others show field dependence. Basically, a field-independent person approaches learning *analytically*, mentally putting together parts to make a whole. The field-dependent person tends to process learning *globally*, viewing each situation as a whole. Speer points out that mathematics textbooks often require analytical (field-independent) skills. On the other hand, a field-dependent student might thrive in open-ended, discovery learning situations (19).

- The behavior of some children tends to be naturally *impulsive*. On the other hand, there are children who are generally *reflective* by nature. Since many primary grade activities, like basic facts drill, favor impulsive behavior, there is a danger that more reflective students will be labelled as "slow learners" (19).

There is evidence, from a study with second and third graders, that reflectivity increases with age (5). This research also suggests that reflective students tend to have higher mathematics achievement.

- Some children favor reasoning *inductively*, or by similarities ("This 5 by 6 rectangle has an area of 30, the 4 by 6 rectangle has an area of 24, and the 3 by 5 rectangle has an area of 15. It looks like the rule for area is 'length times width gives area.' "). Others are inclined to reason *deductively* ("The area of this 5 by 6 rectangle is 30 because there are 5 columns with 6 square units in each. So you can always get the area of a rectangle by multiplying length times width.") (19).

- Children vary in their *tolerance for ambiguity*. This may not always be a factor in the learning of mathematics, but it can occasionally be troublesome for some children. Speer (19) points out that phrases like "reducing fractions," where the word "reducing" does not imply a lessening of the size of the fraction but a change to lower terms, can be very confusing to students who have a low tolerance for ambiguity.

No matter how sharp teachers are in noticing learning styles among students, there will be little good for the students if it all leads only to labelling. What, then, can be done to accommodate different learning styles? Much can be done, but *flexibility* and *variety* are crucial. Here are a few suggestions:

- Provide open-ended problem-solving experiences that foster independence, but also provide guidance in problem-solving techniques. One writer has suggested introducing anxious children to independent experiences gradually, in situations where the topics are not new to the children (21). As an example, once a topic, say long division, has been introduced and practiced, a teacher can work with students in developing a sense of reasonableness for long division exercises. ("In the first problem below, 41 is not a reasonable answer because 40 is close to 41, 20 × 40 = 800, and that isn't close to 8020. How could you convince someone that 52 is not a reasonable answer in the second problem?")

$$20\overline{)8020}^{\,41} \qquad 30\overline{)930}^{\,52}$$

- Encourage a variety of thinking in problem-solving situations. While reflective children might have an advantage in thinking through a problem to a solution, more impulsive children can be in the forefront of suggesting alternative strategies and different problems, if they are given the chance. ("I'm thinking of a number. When you multiply the number by 2 and then add 1, you get 11. What's the number? That's right, 5. Can you make up a problem like that, Jim?")

- Finally, make a variety of materials available to the children at all times. Research implies that *all* students can benefit from exposure to multiple representations of the same concept. (See reference 20 and the *Research Within Reach* bulletins "The Bridge from Concrete to Abstract" and "The Role of Manipulatives.") Learning styles, however, add another dimension to the importance of varying manipulatives. There is evidence that some children favor *continuous* materials and others *discontinuous* materials (18). For example, one child will grasp number concepts easily with continuous materials which make demands on spatial perception, like Cuisenaire rods—the orange rod represents ten, the yellow rod five, and so on. A second child might gain a clearer understanding of number by handling discontinuous materials like poker chips.

A variety of available materials can open doors to understanding mathematics for children who are used to being excluded. In an experiment in teaching geometry concepts, like square, angle, and tetrahedron, to third-graders, Prigge compared the performance of three groups: one which worked with no manipulative aids, another which worked with two-dimensional manipulatives (such as paper-folded squares and rectangles), and a third group which worked with solid geometric manipulatives (such as cubes, tetrahedrons, and clay-formed solids). Among students who had shown up as low-ability on the Iowa Test of Basic Skills, the opportunity to practice with solid geometric manipulatives was especially rewarding. They performed better than their counterparts from the other two approaches (non-manipulative and two-dimensional manipulative) on tests that measured the learning and retention of geometric principles and the ability to transfer those principles to problem-solving situations (15).

Individualizing Instruction

Besides diagnosing learning styles and providing a variety of materials, what can teachers do to accommodate individual differences among their students? Here are a few suggestions:

The *amount of practice* assigned by teachers should vary. One study (9) compared the long-term retention of mathematical skills among three groups of students—one consisting of students who were free to choose their amount of practice, a second consisting of students whose optimal amounts of practice had been individually determined beforehand, and a third group which received a fixed amount of practice. The fixed-practice group compared unfavorably with both the free-choice group and the individualized practice group in a measure of long-term retention.

Suggestions for individualizing practice can be found in the *Research Within Reach* bulletin "Securing Mathematical Skills: Drill and Other Topics" and in references 4, 6, 7, 11, and 13. However, we have already alluded to one very important factor when we talked about the BTES results. At all levels of elementary school, but especially at the primary level, children need to experience more than 50 percent of their practice as *highly successful* (1, 8, 16). Children in the BTES study who spent more time than the average in high-success activities had higher achievement scores in the spring, better retention of learning over the summer, and more positive attitudes toward school (8). The teacher's role in ensuring successful practice begins with a careful diagnosis of each child's strong and weak areas. At the end of the practice, the teacher should make sure that each child both acknowledges and takes credit for the success (8, 16).

If at least half of children's practice is planned for high success, then they can profit from additional practice that challenges their ability to reach problem solutions. As an example, if a group of children has a firm hold on adding multidigit numbers, then their teacher might ask them to find the digits that are missing from the following addition problem:

$$\begin{array}{r} 653 \\ 9\square 2 \\ 87\square \\ \hline 2474 \end{array}$$

This bulletin should be read in conjunction with the *Research Within Reach* bulletin "Grouping for Elementary School Mathematics," because small-group work offers an excellent framework for individualizing instruction. For example, four or five children who learn best aurally can be grouped for basic facts practice that stresses hearing.

Problem solving is another area where small-group work can increase individual attention. Problem solving is a complex process, and it usually happens that each child, with a unique mixture of strong points and weak points, can contribute something special to the process. When small groups are used with whole-class instruction, there is the basis for an environment in which problem-solving strengths will be highlighted and more efficiently used and in which confidence will grow as weaker skills are worked on. As an example, small groups can provide an environment in which anxious children feel safer in embarking on discovery experiences (11, 21).

Some established practices and programs provide further ideas on individualizing instruction. For example, the laboratory approach to mathematics teaching combines individual work with small-group work, involves extensive work with concrete materials, and provides ample opportunity for group and individual discovery experiences (4, 13).

The Developing Mathematical Processes (DMP) program, developed by the Wisconsin R&D Center for Individualized Learning and designed for grades K-6, uses linear measurement as an early and systematic vehicle for developing skills such as comparing, classifying, and ordering. The program continues to use student measurement, especially hands-on measurement, of length, weight, angles, and so on, as a way of developing more sophisticated mathematical concepts and skills (22).

Another K-6 program, which uses a systematic blend of whole-class instruction and individual work to manage individual differences among children, is the Comprehensive School Mathematics Program (CSMP), developed by CEMREL, Inc. Using a combination of whole-class demonstration and discussion, mathematical games, and regular individual work based on a set of exercise booklets that are graded for difficulty, CSMP emphasizes problem solving experiences as the foundation of mathematical understanding (23).

Conclusion

How a teacher should individualize mathematics instruction depends very much on the individuals being taught. Flexibility, observation, and variety must be the watchwords for anyone sensitive to individual styles and needs. But research can help. In the continuing effort to decipher the process of learning mathematics, some beautiful examples keep emerging that testify to the rich potential within every child's mind. For example, the recent study of Carpenter, Hiebert, and Moser demonstrated that many entering first graders, with *no* formal training in the concepts of addition and subtraction, have developed their own understanding of those concepts in a very individual and spontaneous fashion and are able to use that understanding to solve word problems (2, 3).

Sadly, those individually developed skills and concepts can be lost in the rigid instruction in addition and subtraction that often occurs in later grades. Individualizing instruction is a challenging task for teachers, but the price for not attempting it is very, very high.

References

1. Beginning Teacher Evaluation Study (BTES). *Final Report.* San Francisco, California: Far West Laboratory for Educational Research and Development, 1978. (1855 Folsom Street, San Francisco, CA 94103)

* 2. Carpenter, Thomas P.; James Hiebert; and James M. Moser. *The Effect of Problem Structure on First Graders' Initial Solution Processes for Simple Addition and Subtraction Problems.* Madison, Wisconsin: Wisconsin R&D Center for Individualized Learning, 1979.

3. Carpenter, Thomas P., and James M. Moser. *An Investigation of the Learning of Addition and Subtraction.* Madison, Wisconsin: Wisconsin R&D Center for Individualized Learning, 1979.

* 4. Cathcart, W. George (ed.) *The Mathematics Laboratory: Readings from the Arithmetic Teacher.* Reston, Virginia: National Council of Teachers of Mathematics (NCTM), 1977.

5. Cathcart, W. George, and Werner Liedtke. "Reflectiveness/Impulsiveness and Mathematics Achievement." *Arithmetic Teacher,* 16 (November 1969): 563-567.

* 6. Crosswhite, F. Joe, and Robert E. Reys (eds.) *Organizing for Mathematics Instruction.* 1977 NCTM Yearbook. Reston, Virginia: NCTM, 1977.

7. Dunn, R.D., and K. Dunn. *Teaching Students through Their Individual Learning Styles: A Practical Approach.* Reston, Virginia: Reston Publishing Company, Inc., 1978.

8. Filby, Nikola. "Beginning Teacher Evaluation Study: An Overview and Future Activities." Sacramento, California: Commission on Teacher Preparation and Licensing, October 1978 (Newsletter).

9. Gay, Lorraine R. "Use of a Retention Index for Mathematics Instruction." *Journal of Educational Psychology,* 63 (October 1972): 466-72.

10. Hopkins, Martha H. "The Diagnosis of Learning Styles in Mathematics." *Arithmetic Teacher,* 25 (April 1970): 47-50.

*11. Junge, Charlotte W. "Adjustment of Instruction (Elementary School)." In *The Slow Learner in Mathematics,* 35th NCTM Yearbook. Reston, Virginia: NCTM, 1972.

12. Kamii, Constance K. "Evaluation of Learning in Preschool Education: Socio-Emotional, Perceptual Motor, and Cognitive Development." In Benjamin S. Bloom, et al. (eds.), *Handbook on Formative and Summative Evaluation.* New York: McGraw-Hill, 1971.

*13. Mathematics Resource Project. "Teaching Via Lab Approaches." In *Didactics and Mathematics.* Palo Alto, California: Creative Publications, Inc., 1978.

14. National Advisory Committee on Mathematical Education. *Overview and Analysis of School Mathematics: Grades K-12.* Reston, Virginia: NCTM, 1975.

15. Prigge, Glenn R. "The Differential Effects of the Use of Manipulative Aids on the Learning of Geometric Concepts by Elementary School Children." *Journal for Research in Mathematics Education,* 9 (November 1978): 361-367.

16. Schneider, E. Joseph. "Researchers Discover Formula for Success in Student Learning." *Educational R&D Report,* 2 (Fall 1979): 1-6.

*17. Schoen, H.L. "Self-Paced Mathematics Instruction: How Effective Has It Been?" *Arithmetic Teacher,* 23 (February 1976): 90-96.

18. Sharma, Mahesh. "The Problem of the 'Missing Addend.'" *Math Notebook.* Framingham, Massachusetts: The Center for Teaching/Learning of Mathematics, 1979.

*19. Speer, William R. "'Do you See What I Hear?' A Look at Individual Learning Styles." *Arithmetic Teacher,* 27 (November 1979): 22-27.

20. Suydam, Marilyn N., and Jon L. Higgins. *Activity-Based Learning in Elementary School Mathematics: Recommendations from Research.* Reston, Virginia: NCTM, 1977.

21. Trown, Anne. "Teaching Style, Mathematics and Children." *Mathematics Teaching.* 82 (March 1978): 29-32.

22. Wisconsin R&D Center for Individualized Learning. *Developing Mathematical Processes.*

References (continued)

Madison, Wisconsin. For more information on this program, write Developing Mathematical Processes, P.O. Box 7600, Chicago, IL 60680.

23. Woodward, Ernest. "CSMP: A New Alternative in Elementary School Mathematics." *Arithmetic Teacher, 27* (February 1980): 20-27. More information on CSMP can be obtained from CEMREL, Inc., 3120 59th Street, St. Louis, MO 63139.

*These are the readings we believe would be especially interesting to and useful for teachers.

Research Within Reach
Elementary School Mathematics

Algorithms in Elementary School Mathematics

Question: Subtracting multidigit numbers seems to be a difficult skill for many children to learn. Are there ways to make it less difficult?

In order to solve mathematical problems, the mind must be able to concentrate on problem-solving processes. This means that basic facts and computational procedures must be established and readily available, because time spent on hunting down facts, or on reconstructing procedures that should be familiar, is time wasted, time that should be devoted to problem solving. For example, if a child is asked how many pennies will be left after 1000 pennies have been distributed in equal shares among 12 children, then the child should be able to focus on analyzing the problem and on setting up an appropriate computation. Once a computational procedure has been chosen—probably long division, in this example—then the child should be able to proceed through the problem, paying little attention to the principles that support the actual computational steps.

The routine, step-by-step procedures used in mathematical computation are called algorithms. The teaching of algorithms demands care, because their power as mathematical tools depends on the quality of the children's understanding and the quality of the practice that teachers make available. In this bulletin, we look at what research has to say to teachers about choosing, developing, teaching, and practicing algorithms with their students.

The Role of Algorithms

Algorithms are tools, not educational ends in themselves, and they have a proper role in the curriculum as long as they contribute to effective problem solving and to the understanding of mathematical applications.

On the other hand, if teachers try to give all

of mathematics a routine, algorithmic flavor, then learning suffers. In the recent National Assessment of Educational Progress (NAEP) in mathematics, word problems were especially troublesome for students of all ages (8). In their interpretation of the results, a panel of researchers indicted mechanistic, algorithmic approaches to problem solving—for example, urging children to look for key words in problems and to respond invariably to those words ("subtract when you see the word 'less' "). As the panel noted, many textbooks invite mechanistic approaches to problem solving by filling their pages with nearly identical problems, thus displaying little regard for the importance of flexibility in mathematical thinking (20). As a result of not developing flexible problem-solving strategies, "students appear to be learning many mathematical skills at a rote manipulation level and do not understand the concepts underlying the computation" (8, p. 47).

While mechanistic, algorithmic procedures are inappropriate for higher thinking processes like problem solving, they fit naturally with the development of computational skills, such as whole-number addition, subtraction, multiplication, and division, and the corresponding operations for fractions and decimals. As children face the challenge of learning a growing body of mathematical concepts, it is essential that such algorithms be firmly established and maintained through meaningful instruction. Teachers must aim for the *timely* teaching of algorithms, paced to fit with each child's cognitive development and tied closely to each child's understanding of the mathematical meaning behind the algorithmic steps.

Developing Algorithms

Algorithms must develop in the same way that the child's ability to understand mathematics develops, through stages that flow from concrete representations through pictorial representations, and, finally, to symbolic representations. The 1978 yearbook of the National Council of Teachers of Mathematics provides detailed sequences for developing algorithms for whole-number addition, subtraction, multiplication, and division (26). The articles there by Hazekamp and Merseth (13, 19) must be read for a complete description of the suggested sequences. Here, however, is a brief outline of the sequence for developing the addition algorithm (19). In the concrete stages, base-ten blocks (pictured below) are used, but other models, such as beansticks or Cuisenaire rods, would also be appropriate:

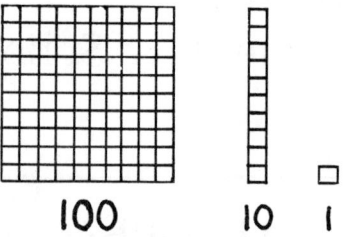

The pre-addition activities of the sequence stress familiarity with the blocks and provide games in which children use the blocks to represent numbers.

At the next level, the children learn to use the special columned design of a playing board to record numbers in a concrete-pictorial fashion.

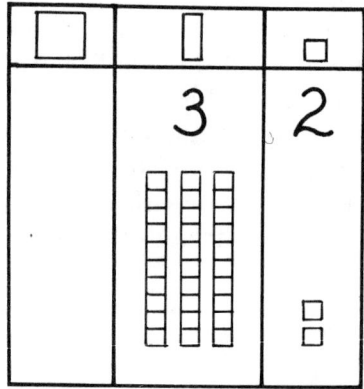

When the children are comfortable using the blocks with their playing boards to represent numbers, the teacher introduces addition problems orally. For the problem, "Build eleven on your board. Now build twenty-four. How many blocks are there altogether?" the children can be guided from representation of 11 + 24 (A) to the middle stage (B) and, eventually, to the more symbolic stage (C). The rate at which a child can progress through the stages depends to a large extent on his or her developmental level and readiness to understand symbolic representations.

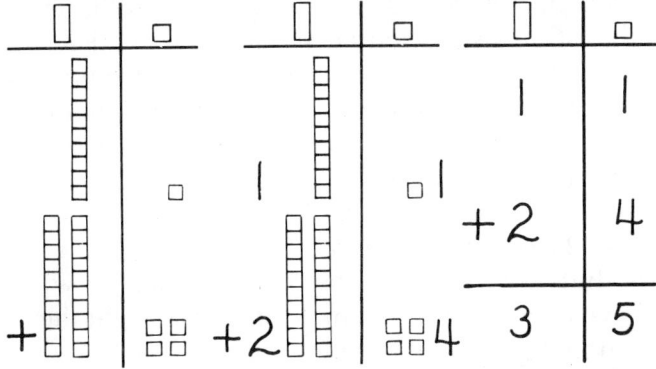

In a discussion about children's use of defective algorithms, Ashlock wrote that "teachers who introduce paper-and-pencil procedures while a child still needs to work problems out with concrete aids are encouraging the child to try to memorize a complex sequence of mechanical acts. This, again, prompts the child to adopt simplistic procedures he can remember" (1, p. 6).

Teaching sequences like the one described above provide a framework in which each child can develop an understanding of algorithms at a pace consistent with individual development. Both National Assessments of Educational Progress (NAEP) in mathematics have provided ample evidence that most nine year olds are still learning the computational algorithms they have been practicing in school (7, 8). Patience and continuous maintenance must mark the application of any teaching sequence for algorithms.

Teaching Algorithms with Meaning

When an algorithm has been learned, its value depends in good measure on how smoothly a user can proceed through its steps without attending to the principles that make it work. The process of *teaching* the algorithm, however, is a different story. Research is clear that children learn an algorithm better when they are given the opportunity to see the sense that connects its steps. Presentations of algorithms as mechanical recipes without meaning short-change students, especially when the students will later be expected to remember an algorithm or transfer their understanding to the development of other algorithms (25).

Teaching algorithms by rote makes no sense at all in light of the research on concept-formation. The exact nature of concept formation is not yet clear to researchers, but it is evident from several studies that very young children develop, in spontaneous fashion, their own conceptual understanding of addition, subtraction, multiplication, and division. In fact, many children are able to solve addition and subtraction word problems even before they enter the first grade (9). Those homegrown concepts are fragile, however, and need nurturing. Since premature exposure to symbolic algorithms can severely cripple conceptual understanding, it is imperative that the concepts be nurtured through meaningful instruction. An excellent guide that makes use of classroom transcripts to alert teachers to the opportunities that arise for this nurturing is the recent book by Brandau and Easley (4).

Teaching for meaning in mathematics is an accepted prescription from research that extends to the entire curriculum, not just algorithms. The *Research Within Reach* bulletin, "Meaning in Elementary School Mathematics," covers these broader aspects of meaning. In the case of algorithms, it is important that children see the sense in each step and relate that sense to their use of the algorithm on the concrete and pictorial levels.

It is also essential that children have a strong grasp on prerequisite concepts. Ashlock has written, "Probably the chief reason children have difficulty with algorithms for whole-number operations is that they do not have an adequate understanding of multi-digit numerals at the time they are introduced to the algorithms" (1, p. 140). He goes on to say that a child who can represent a given number with manipulatives *and* who can go the other way, giving the number name for a given concrete representation, probably understands multi-digit numeration and is ready for the meaningful development of algorithms.

In later grades, when the focus turns to computation with fractions, prerequisite concepts are also critical. The first NAEP assessment left the clear impression that "many students of junior high school age have little computational skill with fractions and probably little conceptual understanding" (6). In the recent NAEP assessment, the lack of conceptual understanding of fractions showed up again, as did the dependency of many students on defective algorithms. In the exercise ¾ + ½ = ? 34 percent of the thirteen year olds and 15 percent of the seventeen year olds added numerators and added denominators to produce the incorrect answer 4/6. Yet the assessment results showed that when students were using the correct addition algorithm for fractions, the complexity of the denominators had little effect on their performance, a phenomenon which prompted the following observation: "Apparently if students have learned a computational algorithm they can apply it successfully in most situations, but if they have not mastered the algorithm they cannot reconstruct it or fall back on intuitive models to solve problems" (8, p. 12).

The researchers who interpreted the results of both assessments pointed to the lack of conceptual understanding as a major cause of the weak computational skills and urged that children get a more extensive exposure to fraction concepts, *before* teachers introduce algorithms for computing with fractions. In particular, children should be clear about units, part-whole relations, comparing and ordering fractions, and renaming mixed and improper fractions, and they should know the vocabulary and symbolism of fractions. Furthermore, they should be able to answer questions like "Which fraction is greater?" or "How much is shaded in this diagram?" before they tackle the algorithms (6).

As an example of the benefits to computation that a sound conceptual foundation in fractions can have, consider the algorithm for fraction multiplication ("Multiply numerators and multiply denominators: 2/3 × 4/5 = 8/15"). The algorithm's simplicity disguises a sophisticated concept, and research recommends developing the algorithm with the concepts of rectangular area and fractional parts (12). For example, once children can recognize from visual evidence that a 6 cm. by 4 cm. rectangle has an area of 6 × 4 = 24 sq. cm., then it is possible for them to deduce that a ½ cm. by ½ cm. rectangle will have an area of ¼ sq. cm. From that insight they can make sense out of the equation ½ × ½ = ¼.

For further ideas about developing computational skills with fractions, see references 5, 11, 18, and 25 and the *Research Within Reach* bulletin "Sequence With Substance: The Elementary School Mathematics Curriculum."

Beyond Paper and Pencil

It no longer makes sense to teach algorithms only as paper-and-pencil exercises. Research studies on mental arithmetic, on estimation skills, and on calculator use clearly underscore that conclusion.

From their early years on, students should be taught mental arithmetic skills and encouraged on a regular basis to use them. One such skill is *compensation*, which might inspire the following kind of thinking:

> 6 × 17 = ? *I know that* 6 × 20 = 120, *and that* 6 *threes are* 18, *so, since* 17 = 20 − 3, 6 × 17 *must equal* 120 − 18, *or* 102.

Also, from the primary grades on, children should be taught how to estimate and should be *regularly* asked questions that put them in the habit of checking to see if their answers are reasonable. If you as teacher assign an exercise like 259 + 723, preface the written work with questions like, "How big do you think the answer will be? About 1000? What makes you think that?" Then, at the end of the written work, ask, "Does your answer seem reasonable? Is it too big? Too small?" For suggestions on teaching children to do mental arithmetic, to estimate, and to decide whether answers are reasonable, see reference 32 and the *Research Within Reach* bulletin, "Estimation and Mental Arithmetic."

Computational algorithms, estimation, reasonableness, and mental arithmetic make up a versatile set of tools for learning mathematics. An excellent vehicle for keeping those tools working together is the hand-held calculator. Contrary to some fears and opinions, calculators will not eliminate the need for children to learn about algorithms. They may eliminate the need to use traditional paper-and-pencil algorithms on a regular basis, but they need not, and in fact *cannot*, replace conceptual understanding and computational skill. In fact, as calculator proficiency and use increase, students will need different, nontraditional algorithms to aid them in solving problems. For example, in their quest for a final solution to a problem, calculator users must often employ an algorithm, or algorithms, for recording intermediate answers.

The recent NAEP assessment provides a good illustration of the challenges that attend calculator use. Several exercises given to thirteen year olds were assigned both to children with calculators and to children without calculators. On problems involving direct, one-step division with whole-number answers, calculator users had, in general, a higher success rate than nonusers. However, on problems like the one below, where there is a division remainder, non-users had the better record (22).

A man has 1,310 baseballs to pack in boxes which hold 24 baseballs each. How many baseballs will be left over after the man has filled as many as he can?

In order to succeed at such problems, children must understand the concept of division, must know when remainders occur as well as how they can be determined, and then must interpret the results. In effect, they must be able to understand, but make a different use of, the long-division algorithm.

Used regularly in conjunction with algorithm instruction, the calculator can be a powerful tool for developing and reinforcing computation skills, estimation skills, and problem-solving skills. The growing body of research on calculator use confirms this judgment, as detailed in references 3 and 31 and the *Research Within Reach* bulletin, "Calculators in the Classroom."

Comparing Algorithms for Whole Numbers

When two different procedures can be used to get the same computational result, which algorithm should teachers choose? The comparison of algorithms has been the focus of many research studies, and in this section we list some of the major results.

Subtraction of whole numbers. Most of the research attention in subtraction instruction has been on the *equal additions* approach and the widely used *decomposition* approach.

$$\begin{array}{r} 3\overset{1}{6}7 \\ -1\overset{}{4}9 \\ \hline 218 \end{array} \qquad \begin{array}{r} 3\overset{5}{\cancel{6}}\overset{1}{7} \\ -149 \\ \hline 218 \end{array}$$

Equal Additions   Decomposition

In comparing the two approaches, researchers have determined that decomposition is superior for both understanding and accuracy (25).

Addition of whole numbers. Wheatley compared two methods of column addition, one in which the numbers are added together in successive fashion (in the example below, successive addition is illustrated by "4 + 7 = 11, 11 + 6 = 17, 17 + 3 = 20, 20 + 5 = 25"), and another in which pairs of numbers whose sum is 10 are matched first and then combined (in the same example, "3 and 7 are 10, 6 and 4 are 10, so the column adds up to 10 + 10 + 5 = 25").

$$\begin{array}{r}4\\7\\6\\3\\\underline{5}\end{array}$$

The study revealed that the former method, the *direct* method, was faster than the *tens* method, but the methods were comparable in accuracy (29, 30).

Division of whole numbers. A study comparing the *distributive* and *subtractive* methods among third graders implied that there are advantages to both algorithms. Low-ability students showed a better understanding of division with the subtractive method, while children taught the distributive method achieved higher problem-solving scores (25, 27).

$$\begin{array}{r}21\\24\overline{)504}\\\underline{48}\\24\\\underline{24}\\0\end{array} \qquad \begin{array}{r}21\\24\overline{)504}\\\underline{-240}\quad 10\times 24\\264\\\underline{-240}\quad 10\times 24\\24\\\underline{-24}\quad +\;1\times 24\\0\qquad 21\end{array}$$

Distributive   Subtractive

If each division algorithm has its advantages, then a question arises: Should teachers expose their students to more than one algorithm for the same operation? The answer is not clear. In a third-grade, long-division study, children who were taught both methods increased their understanding of the concept of division (24). On the other hand, another study, which involved the learning of algorithms for translating numerals from base 10 to other bases, provided evidence that the introduction of two algorithms in a short amount of time can cause interference in student understanding (2).

From the research data available at the present time, it is impossible to provide final answers on the use of alternative algorithms. Several recommendations can be made, however.

If a second, alternative algorithm is introduced to students, then there should be a thorough discussion of how the two algorithms are similar and how they are different. All appropriate connections between the underlying concepts should be emphasized.

When two algorithms offer separate advantages for the same operation, then choice should hinge on each student's individual needs and individual style. Once the choice has been made for a particular student, however, the teacher's attention and energy must focus on securing the algorithm through regular practice. According to research, short (5 to 10 minute), daily periods of practice are more effective than staying with an algorithm until everyone gets it (18).

For further information on practice and mastery, see the *Research Within Reach* bulletins, "Securing Mathematical Skills: Drill and Other Topics" and "Evaluation in Mathematics Education. Part Two: Mastery Learning in Elementary School Mathematics."

Algorithms and Computational Difficulties

Children have the best chance to learn an algorithm when it has been developed with meaning. Research also suggests that *students should practice on algorithms only when they know the basic arithmetic facts at a stimulus/response level.*

The research on systematic computational errors makes it clear that many students have gone astray in their understanding of algorithms (1, 10, 17). Indeed, for a large number of students it is very difficult to overcome the anxiety created by their consistent computational failure. Unfortunately, students frequently deal with this anxiety and the attendant discouragement by effectively withdrawing from the pursuit of mathematical learning.

For teachers of mathematics, this computational anxiety presents a special challenge. Several sources provide alternative approaches for the remedial teaching of algorithms: small-group practice (16), ways to remove undue stress on "the correct answer" (1), alternative "low-stress" algorithms (14), and the use of flow charts or diagrams in the teaching of algorithms (28). Ashlock's book, in particular, provides a number of examples, both of typical systematic errors and of ways to eliminate them (1). Finally, calculators can enable students to develop and use new algorithms and to gain a fresh perspective on traditional algorithms (3).

Conclusion

With teachers attending to individual styles and needs, it is possible for *all* children to learn algorithmic techniques. Computation can and should take its appropriate place in mathematics instruction—as a *tool* to be used in solving problems and in applying mathematics to the real world.

References

1. Ashlock, Robert B. *Error Patterns in Computation: A Semi-Programmed Approach.* Columbus, Ohio: Charles E. Merrill Publishing Co., 1976.
2. Barszcz, Edward L., and J. Ronald Gentile. "Retroactive Interference of Similar Methods to Teach Translation of Base Systems in Mathematics." *Journal for Research in Mathematics Education,* 7 (May 1976): 176-182.
3. Beardslee, Edward C. "Teaching Computational Skills with a Calculator." In Marilyn N. Suydam and Robert E. Reys (eds.), *Developing Computational Skills,* 1978 NCTM Yearbook. Reston, Virginia: National Council of Teachers of Mathematics (NCTM), 1978.
4. Brandau, Linda, and Jack Easley. *Understanding the Realities of Problem Solving in Elementary School: With Practical Pointers for Teachers.* Columbus, Ohio: ERIC Clearinghouse for Science, Mathematics, and Environmental Education (ERIC/SMEAC), 1979.
5. Capps, L.R. "A Comparison of the Common Denominator and Inversion Method in Teaching Division of Fractions." *Journal of Educational Research,* 56 (July/August 1963): 516-522.
6. Carpenter, Thomas P., et al. "Notes from National Assessment: Addition and Multiplication with Fractions." *The Arithmetic Teacher,* 23 (February 1976): 137-143.
7. Carpenter, Thomas P., et al. *Results from the First Mathematics Assessment of the National Assessment of Educational Progress.* Reston, Virginia: NCTM, 1978.
8. Carpenter, Thomas P., et al. "Results and Implications of the Second NAEP Mathematics Assessment: Elementary School." *The Arithmetic Teacher,* 27 (April 1980): 10-14, ff.
9. Carpenter, Thomas P., James Hiebert, and James Moser. "The Effect of Problem Structure on First-Graders' Initial Solution Processes for Simple Addition and Subtraction Problems." Madison, Wisconsin: Wisconsin R&D Center for Individualized Schooling, 1979.
10. Cox, L.S. "Systematic Errors in the Four Vertical Algorithms in Normal and Handicapped Populations." *Journal for Research in Mathematics Education,* 6 (November 1975): 202-221.
11. Ellerbruch, Larry W., and Joseph N. Payne. "A Teaching Sequence from Initial Fraction Concepts through the Addition of Unlike Fractions." In M.N. Suydam and R.E. Reys (eds.), *Developing Computational Skills,* 1978 NCTM Yearbook. Reston, Virginia: NCTM, 1978.
12. Green, Geraldine Ann. "A Comparison of Two Approaches and Two Instructional Materials on Multiplication of Fractional Numbers." Unpublished Ph.D. dissertation, University of Michigan, 1969.
13. Hazekamp, Donald W. "Teaching Multiplication and Division Algorithms." In M.N. Suydam and R. E. Reys (eds.), *Developing Computational Skills,* 1978 NCTM Yearbook. Reston, Virginia: NCTM, 1978.
14. Hutchings, Barton. "Low-Stress Algorithms." In Doyal Nelson and Robert E. Reys (eds.), *Measurement in School Mathematics,* 1976 NCTM Yearbook. Reston, Virginia: NCTM, 1976.
15. Hynes, Michael C. "Using Manipulative Aids to Model Algorithms in Remedial Situations." In

References (continued)

Mary Ellen Hynes (ed.), *Topics Related to Diagnosis in Mathematics for Classroom Teachers*. Bowling Green, Ohio: Research Council for Diagnostic and Prescriptive Mathematics, 1979.

16. Junge, Charlotte W. "Adjustment of Instruction (Elementary School)." In *The Slow Learner in Mathematics*, 35th NCTM Yearbook. Reston, Virginia: NCTM, 1972.

17. Lankford, Francis G., Jr. "What Can a Teacher Learn about a Pupil's Thinking through Oral Interviews?" *The Arithmetic Teacher*, 21 (January 1974): 26-32.

18. Mathematics Resource Project. "The Teaching of Skills." In *Didactics and Mathematics*. Palo Alto, California: Creative Publications, Inc., 1978.

19. Merseth, Katherine Klippert. "Using Materials and Activities in Teaching Addition and Subtraction Algorithms." In M.N. Suydam and R.E. Reys (eds.), *Developing Computational Skills*, 1978 NCTM Yearbook. Reston, Virginia: NCTM, 1978.

20. National Assessment of Educational Progress. *The Second Assessment of Mathematics, 1977-78: Mathematical Applications*. Denver, Colorado: Education Commission of the States, 1979.

21. National Assessment of Educational Progress. *The Second Assessment of Mathematics, 1977-78: Mathematical Knowledge and Skills*. Denver, Colorado: Education Commission of the States, 1979.

22. National Assessment of Educational Progress. *The Second Assessment of Mathematics, 1977-78: Released Exercise Set*. Denver, Colorado: Education Commission of the States, 1979.

23. Roberts, Gerhard H. "The Failure Strategies of Third Grade Arithmetic Pupils." *The Arithmetic Teacher*, 15 (May 1968): 442-446.

24. Scott, A. "A Study of Teaching Division through the Use of Two Algorithms." *School Science and Mathematics*, 63 (December 1963): 739-752.

25. Suydam, Marilyn N., and Donald J. Dessart. *Classroom Ideas from Research on Computational Skills*. Reston, Virginia: NCTM, 1976.

26. Suydam, Marilyn N., and Robert E. Reys (eds.). *Developing Computational Skills*, 1978 NCTM Yearbook. Reston, Virginia: NCTM, 1978.

27. Van Engen, Henry, and E. Glenadine Gibb. *General Mental Functions Associated with Division*. Educational Service Studies, No. 2. Cedar Falls, Iowa: Iowa State Teachers College, 1956.

28. Wheatley, Grayson H. "Mathematical Road Maps: A Teaching Technique." *The Arithmetic Teacher*, 23 (January 1976): 18-20.

29. Wheatley, Grayson H. "A Comparison of Two Methods of Column Addition." *Journal for Research in Mathematics Education*, 7 (May 1976): 145-154.

30. Wheatley, Grayson H., and Charlotte L. Wheatley. "How Shall We Teach Column Addition? Some Evidence." *The Arithmetic Teacher*, 25 (January 1978): 18-20.

31. Wheatley, Grayson H., et al. "Calculators in Elementary Schools." *The Arithmetic Teacher*, 27 (September 1979): 18-21.

32. Zepp, Raymond. "Algorithms and Mental Computation." In Marilyn N. Suydam and Alan R. Osborne (eds.), *Algorithmic Learning*. Columbus, Ohio: ERIC/SMEAC, 1976. (ED 113 152)

Stronger Curriculum

Research Within Reach
Elementary School Mathematics

Mathematical Problem Solving: Not Just a Matter of Words

Question: My third graders score in the satisfactory to high range both in reading and computation. However, they are consistently low in solving word problems successfully. What insight or training can I give them to improve their problem solving skills?

Adult responses to the phrase "problem solving" may vary, but for many people it triggers memories, often uncomfortable, of textbook word problems. This misconceived notion—that mathematical problem solving and word problems are one and the same—is a common one. It pervades the classroom strategies of many elementary school teachers. Word problems are included in this bulletin, but only as one facet of the development of children as they learn to recognize mathematical structures behind problem situations and as they begin to use mathematical strategies to reach problem solutions.

In studying mathematical problem solving, researchers have looked carefully at the makeup of problems, at the strategies used by good problem solvers, and at the behavior of teachers when they teach problem solving. This report focuses on their observations, with emphasis on the elementary school level.

Problem Selection

- Provide your students with non-textbook problems and problem settings. The end-of-the-chapter exercises that often come equipped with recipes for solving problems are very limited in the problem solving growth they offer your students. These problems should be used, but they should be complemented with teacher-written problems. Furthermore, research suggests that you should invite regular involvement of students in the composition of problems (8). Children derive more enjoyment from original problems, and the process of formulating problems can help children to focus on important details.

As an example of a way to introduce original problems, suppose the text has a series of exercises like the following: "John, Barbara, and Tom went clothes shopping with $18.75. How much will each get to spend if they divide the money equally?" Once the children have had some practice with the textbook problems, you can expand their awareness if you single out several students and say, "Suppose, Bill, Lucy, Jim, and Maria, that you go to the amusement park with $22.00, but $2.00 has to be set aside for bus fare. How much does each of you get to spend at the park, if you divide the money evenly?" After this problem is discussed and worked through, you can challenge the children to compose similar problems.

- Offer your students frequent opportunities to work with problems that yield a variety of solution avenues. For problem solving skills to develop, flexibility needs room to develop, and too often textbooks present their problems in a manner that suggests there is but one way to solve each type of problem. Multi-solution problems are often difficult to solve, but they form the foundation for lively and profitable instruction in problem solving strategies. In the following example, making a list to search for a pattern and diagram-making are but two strategies toward solution:

    There are 18 students in this class. Suppose each of you plays one game of tennis with every other person in the class. How many games of tennis will be played?

- Make regular use of problems in which computational accuracy does not have the central role. One danger in the development of problem solving skills is that children associate problem solving only with computational accuracy or rote learning (8, 14). Some examples of noncomputational problems follow:

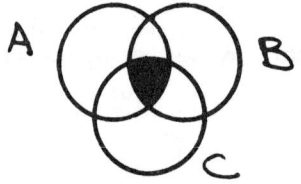

1. The circle marked A contains all numbers from 1 to 50. Circle B contains all odd numbers. Circle C contains all multiples of 7. What's in the shaded region? (1)

2. McKillip and associates developed a program, called "Patterns," for children 3 to 5 years of age. Using blocks or tiles of two colors, the children took on two-fold tasks: to copy patterns made by the teacher and, if possible, to extend the patterns. For example, extend

    Despite their age, the children who participated were able and willing to attend to the problems for long periods of time (10).

    Once introduced to the world of mathematical patterns, children will respond enthusiastically throughout their school careers to the search for patterns, provided patterns become a regular part of their problem solving experience. For more suggestions, see McKillip (10).

3. As hand calculators become more available, they provide a ready source for new mathematical problems. Even richer are the opportunities to present problem solving in a light that doesn't shine just on computational accuracy. As an example, "Using your calculator and any shortcuts you can devise, cross out the numbers that do not belong on the left side of the equal sign:

    $14 \times 3 \times 9 \times 5 \times 8 \times 6 = 3780$"

    Problem solving experiences like this one from Immerzeel (3; also see 12) however, become only so much mechanical magic unless you aid the students, with questions and discussion, in understanding the mathematics that leads to the calculator outcomes. For example, after the above exercise is finished, you could ask, "About what number on the left does the zero in 3780 give you the most information? . . . That's right, the five."

The role that calculators can play in the acquisition of problem solving skills has been the focus of several recent studies. In one sixth-grade study (18), two groups of students received the same training in problem solving processes such as estimating, retracing steps, and checking the reasonableness of answers; however, one group used calculators in the training and the other did not. In a final problem solving test, the calculator group used *significantly* more of these processes than did the members of the noncalculator group. The conclusion: calculators can help students strike a better balance between laboring for computational accuracy and concentrating on problem solving strategies.

- Be flexible in representing problems for your students. Research has shown that problems with materials or with visual aids like diagrams are generally easier for children to deal with than those without such aids (15, 17). Whenever possible, have the children move the materials, make the drawings or diagrams, or even dramatize the problem setting.

Furthermore, if children are given freedom to choose among mathematical models to solve a particular problem, they will select the model that makes the idea most meaningful to them (2). For example, suppose a problem asks, "Bill takes ½ of a candy bar and Sandy takes ¼ of the same candy bar. How much is left?" A child might find it most meaningful to work concretely, folding rectangular strips of paper, or she might choose to diagram the candy bar and its pieces, or she might be most satisfied changing to the symbolic model:

$$2/4 + 1/4 = 3/4; \quad 4/4 - 3/4 = 1/4$$

- Vary the wording in problems of the same type. It is apparently easier for students to work through word problems when numerical data are listed in the order in which they are used (14). Rather than limiting the students' experience to an optimal ordering, however, it would be better for a teacher to change the ordering and wording, perhaps gradually, and to aid the children in relating the wording back to the problem that follows the order of use.

- Vary the settings in problems that require the same mathematical operation. Research indicates that there is less payoff for the student in repeated practice with the same word problems using the same techniques, than there is in solving different types of problems with the same technique (9). For example, the following two problems involving long division would better serve students' needs if they were presented at nearly the same time:

  $13.20 is divided evenly among 11 people. What is each person's share?

  Jane drove her car 648 miles in 12 hours. What was her average speed per hour?

If you were to give your students a series of such problems, it would be important for you to discuss with them the mathematical thread that links the problems together. Otherwise, rote memory might be the students' only recourse for distinguishing among the various problems.

Teaching Problem Solving

As one takes a hard look at problem solving, some qualities of good problem solvers shine through. Seen in the light of the actions problem solvers take, the strategies they try, the opportunities they grab, problem solving appears less veiled in mystery, and teachers can begin to plan their instructional strategies. Therefore, as we list each aspect of good problem solving, we shall pause and make suggestions, some from research, others from the experience of teachers who have had success in teaching problem solving.

One preliminary observation: teacher questions play significant roles in all of the suggestions that follow. If any backbone for the teaching of problem solving can be identified, it is this—the consistent asking of questions that involve students and invite them to think.

Good problem solvers are quick to understand the important features of a problem, in particular, the objective.

Students should be encouraged to put in their own words what they think the problem is about—the information it provides, the information it seeks. With the amusement park problem cited earlier, you could ask questions like, "Sam, where are Bill, Lucy, Jim, and Maria going?" "How much money do they have?" "How are they getting to and from the park, Julie?"

What role does reading skill play in the development of word problem skill? Research does not seem to point to reading difficulty as the most likely culprit when word problem skills are weak. In one sixth-grade study the majority of students who solved problems incorrectly could read the problems orally and even retell the problem stories in their own words (5). Teachers must be prepared to drive deeper than reading comprehension in their teaching of problem solving. Zweng (19) urges teachers to develop in students a language for problem solving, asking frequent questions that are operation-related, such as, "When you put together bunches of the same size, should you multiply or divide?"

When reading difficulties do interfere with students' understanding of word problems, there are several options for the teacher to try. First of all you can vary the type of problem and its representation (diagrams, pictures, and so on), so that reading comprehension is not so critical to solution. Another strategy is to read the word problems aloud or use a tape recorder.

You can also tie together some reading skills training with work on word problems. Two specific suggestions are that you help your students to learn how to select the main ideas when they read word problems, and that you help them to practice making inferences ("Where are the four children going to get the bus fare?") (14).

It is natural to wonder how helpful training aimed at specific words is. One research study associated progress in solving word problems with training in understanding quantitative vocabulary ("one fourth," "the seventh," and so on). However, there is also evidence that "emphasis on *isolated* word clues like 'left,' 'in all,' can be grossly misleading as a problem solving procedure" (15). For example, it would not be very profitable to instruct children that "in all" means you should add or "left" means you should subtract. Consider the following problem:

> Brian spent a certain amount of money on the first ride at the carnival, and twice as much money on each of the four rides that were left. In all, Brian spent nine dollars at the carnival. How much did he spend on the first ride?

Since young children are easily distracted by the presence of irrelevant information in their environment, the elementary school teacher must be carefully attuned to developmental levels. Teachers should take pains to present problems in a way that highlights their critical features, and to minimize extraneous stimuli or information used just to make material interesting—for example, bright background colors (13).

Good problem solvers initiate their approach to problems with less anxiety and with more certainty that they will be able to reach a solution, than do other problem solvers.

This suggests that teachers should group children for concentrated problem solving work according to the levels of difficulty they can tolerate comfortably. Several studies showed that children working on word problems in small groups of three or four, *when given enough time,* solved significantly more problems than those who worked alone (14).

Apparently, then, loosening the stranglehold of time pressure, along with some careful grouping, can do much to relieve problem solving anxiety in children, and to allow them to develop their problem solving skills in a way that is both meaningful and comfortable for them.

Teacher questions can do much to boost student confidence that enough information can be drawn from the wording of a problem for a solution to be reached. ("If each student plays one game of tennis with everyone else in the class, how many games will you be playing, John? What about Betty?")

Good problem solvers can transfer their learning from one solved problem to similar problems.

Three suggestions present themselves for your problem solving instruction. First, *before* the students tackle a problem, ask them if the problem reminds them of any previously solved by the class. If they make some connections, valid or invalid, discuss the connections they have made.

Secondly, *after* the students have successfully solved a problem, help them to reflect on the process that produced its solution. Guiding them by questions through a summary can serve this purpose.

Finally, go one step further and cultivate your students' ability to generalize and see commonalities among problems. ("The tennis game problem you just solved is a lot like the handshake problem you did last week. Do you remember that problem?")

Good problem solvers are able to evaluate and select from alternative solution routes.

You can encourage your students to suggest, out loud, possible strategies for solution. Acknowledge all of the suggestions, keeping a running list and giving all equal attention. Help the students to compare the various suggestions and point out what kind of approach each represents. (In the case of the tennis problem, you might say, "Okay, Mike, you would like to make a list of the people in the class and begin to count pairs to see if a pattern comes up. That's pattern searching. Sarah, you want to make a diagram with two columns, each with 18 dots, and to connect the dots with lines. That's an example of diagram-making," and so on.) For more suggestions in a similar vein, see LeBlanc (8).

Good problem solvers estimate and approximate well, and, when they get solutions, check to see if they are reasonable.

Estimation is a skill covered in some depth in our *Research Within Reach* bulletin "Estimation and Mental Arithmetic." It is a skill that should have a regular role in your mathematics class. However, in your capacity as teacher of problem solving, you will find it an especially timely ally when a child or group becomes bogged down in the search for exact numerical answers. ("About how much will each child get to spend at the amusement park? More than $10?")

Estimating can help problem solvers unravel their thoughts, especially when it is used in conjunction with another technique, problem simplification. At times, problems go unsolved mainly because they involve very large whole numbers or very small fractions, diagrams that have too many lines, or charts that have too many letters. When those situations arise, you should help the students to consider ways to simplify the problem. ("If 18 students are tough to deal with in the tennis problem, pretend there are 4 people or 5 people in the class—that might make things less complicated.")

Provide your students with opportunities to consider and discuss the reasonableness of answers. ("Jane and Mark took an automobile trip. Jane drove 10 7/8 miles, and then Mark drove an additional 15 4/5 miles. About how far did they travel? 25 miles? 27 miles? 30 miles? Which is the most reasonable estimate? Why?")

As you highlight the problem solving process for your students, make it a habit to ask, "Does that answer seem right? What seems right (or wrong) about it?" Ask questions even when the solution your students reach is correct.

Good problem solvers learn from their mistakes, and, when answers do not check out, they are flexible enough to switch to alternative approaches.

At this point you can capitalize on your original listing of the suggested solution strategies. However, before the students launch themselves into a totally new direction, you should guide them in looking at the first approach to see how it might have fallen short. Did it yield an answer far too large? Did its diagram get too complicated? You should help them to consider if they can make adjustments in their first approach before they choose a second approach.

Conclusion

Problem solving in mathematics should not be a part-time endeavor. The kind of teacher questions outlined in the previous sections can be adapted to the entire mathematics program—estimation, checking for reasonableness, learning from mistakes; they all apply to computation, measurement, and mental arithmetic, as well as to word problems. Research has shown, for example, that children who are taught thinking strategies for learning basic addition and multiplication facts retain those facts longer and can use the strategies to learn harder facts (16).

If you apply the strategies outlined in this bulletin throughout your teaching of mathematics, if you highlight problem solving strategies whenever you can, your students will be drawn to the conclusion that problem solving is, indeed, not just a matter of words.

References

1. *CSMP Mathematics for the Intermediate Grades Part III, Teacher's Guide.* St. Louis, Missouri: CEMREL, Inc., Comprehensive School Mathematics Program, 1978.
2. Fennema, Elizabeth H. "Models and Mathematics." *The Arithmetic Teacher,* 19 (1972): 635-640.
3. Immerzeel, George, and Ockenga, Earl. *Calculator Activities for the Classroom.* Palo Alto, California: Creative Publications, Inc., 1977.
4. Kilpatrick, Jeremy. "Research on Problem Solving in Mathematics." *School Science and Mathematics,* 78 (March 1978): 189-192.
5. Knifong, J. Dan, and Boyd D. Holtan. "A Search for Reading Difficulties among Erred Word Problems." *Journal for Research in Mathematics Education,* 8 (May 1977): 227-230.
* 6. Krulik, S. (ed.) *Problem Solving in School Mathematics,* 1980 NCTM Yearbook. Reston, Virginia: National Council of Teachers of Mathematics (NCTM), in press.
7. Krutetskii, V.A. *The Psychology of Mathematical Abilities in Schoolchildren.* Chicago, Illinois: University of Chicago Press, 1976.
8. LeBlanc, John F. "You Can Teach Problem Solving." *The Arithmetic Teacher,* 25 (November 1977): 16-20.
9. Lester, Frank K., Jr. "Ideas about Problem Solving: A Look at Some Psychological Research." *The Arithmetic Teacher,* 25 (November 1977): 12-15.
10. McKillip, William D. "'Patterns'—A Mathematics Unit for Three-and Four-Year Olds." *The Arithmetic Teacher,* 17 (January 1970): 15-18.
*11. Nelson, I. Doyal, and Kirkpatrick, Joan. "Problem Solving." In J.N. Payne (ed.), *Mathematics Learning in Early Childhood,* 37th NCTM Yearbook. Reston, Virginia: NCTM, 1975.
12. Reys, Robert E., et al. *Keystrokes: Calculator Activities for Young Students.* Palo Alto, California: Creative Publications, Inc. 1979.
13. Stevenson, Harold W. "Learning and Cognition." In J.N. Payne (ed.), *Mathematics Learning in Early Childhood,* 37th NCTM Yearbook. Reston, Virginia: NCTM, 1975.
*14. Suydam, Marilyn, and Weaver, J. Fred. *Using Research: A Key to Elementary School Mathematics.* Reston, Virginia: NCTM, 1975.
*15. Suydam, Marilyn, and Weaver, J. Fred. "Research on Problem Solving: Implications for Elementary School Classrooms." *The Arithmetic Teacher,* 25 (November 1977): 40-44.
16. Thornton, Carol A. "Emphasizing Thinking Strategies in Basic Fact Instruction." *Journal for Research in Mathematics Education,* 9 (May 1978): 214-227.
17. Trimmer, Ronald G. "A Review of the Research Relating Problem Solving and Mathematics Achievement to Psychological Variables and Relating These Variables to Methods Involving or Compatible with Self-Correcting Manipulative Mathematics Materials." 1974. [ED 092 402]
18. Wheatley, Charlotte L. *The Effect of Calculator Use on the Problem Solving Strategies of Elementary School Pupils.* Paper presented at the annual meeting of the National Council of Teachers of Mathematics, Boston, April 1979.
19. Zweng, Marilyn H. "The Problem of Solving Story Problems." *The Arithmetic Teacher,* 27 (September 1979): 2-3.

*These are the readings we believe would be especially interesting to and useful for teachers.

Research Within Reach
Elementary School Mathematics

Estimation and Mental Arithmetic

Question: Are there ways to help students appreciate when their mathematical answers are unreasonable? For example, if their work on a soda-buying problem produces a per bottle cost that is much too large, are they suspicious and willing to check for an error?

From office to car to store to home, numbers seem always to be with us. Because the need to estimate or to calculate without pencil and paper permeates our everyday lives, training and practice in those skills should permeate elementary school mathematics.

Among the conclusions drawn from the research and other work with estimation and mental arithmetic, three merit early and special note in this bulletin. First, few schools do *any* estimation work with their students. Second, those schools that do introduce estimation and mental arithmetic generally do not introduce them soon enough. Finally, once they are introduced, the skills are nurtured much too infrequently and inconsistently.

Estimation rarely appears in elementary school textbooks before the third- or fourth-grade level, and, when it does appear, it is often left out of classroom activities. This tradition does not reflect the evidence, from several research studies, that children can benefit from estimation training much earlier than third grade. In fact, even *before* formal mathematics instruction begins, children in kindergarten are able to make correct decisions, without counting, about the relative number sizes of two groups of identical objects (5). They are even able to do some informal reasoning about proportion, such as deciding, after quickly viewing objects, that 12 is closer to 16 than 10 is to 18.

There are some acknowledged prerequisites to formal training in estimation (1):

1. The ability to round a whole number to the nearest 10 or 100.
2. The ability to multiply by powers of 10 (for example, recognizing that 3×40 is the same as 12 tens or 120).
3. The ability to add, subtract, multiply, or divide numbers, each of which is a power of 10 (for example, recognizing that 240 ÷ 30 asks how many times 3 tens divide into 24 tens and, hence, is intimately related to 24 ÷ 3).

Even before these skills take form, teachers can and should help young children to develop a feeling for the approximate location of numbers on the number line. The children can then begin to learn how to round, possibly with the help of unit and ten-unit strips ("Find 68 on the number line. Is it closer to 60 or 70? Which is the nearest ten, 6 tens or 7 tens?"). (See reference 10 for further guidance.)

Estimation is often associated with computational practice ("Is 3 × 35¢ larger than a dollar?"), a relevant activity for all ages. There are other estimation experiences which are also appropriate for young children; for example, estimating associated with counting ("How many marbles do you think there are in this jar?") and measurement estimations ("About how many pennies do you think it would take to cover a dollar bill?"). Armed with estimation skills that they use on a regular basis, children are better prepared to take on work in computation, and their teachers have the opportunity to flavor computational practice with relevant, motivating estimation problems ("How can you tell 4×36 is larger than 120? Because 4×30=120? How do you know that?"). For further examples, see the article by O'Daffer (9).

- Mental arithmetic skills with a sophisticated computational flavor are, of course, inappropriate for very young children. Nevertheless, there are a variety of ways that teachers can lay the foundation for these skills. For instance, teachers should teach thinking strategies to complement concrete activities and drill when they are establishing the basic facts. One example is the "compensation strategy," by which the two numbers being added are changed in compensating fashion to yield a more familiar fact ("9 + 6 = ? 9 and 1 is 10, 10 and 5 more is 15, so 9 + 6 = 15"). Research has shown that instruction in such strategies will improve both the retention of basic facts and the ease with which those facts are used to learn harder facts (11, 14).

- Young children are keen to perceive patterns in mathematics, and skill in mental arithmetic derives in good part from appreciation and exploitation of mathematical regularity. However, because of their level of cognitive development, young children must actually see such regularity. One recommended teaching device is to construct a number chart in rows or columns of 10:

1	2	3	4	5	6	7	8	9	10
11	12	13	14	15	16	17	18	19	20
21	22	23	24	25	26	27	28	29	30
31	32	33	34	35	36	37	38	39	40

Use the chart to have children locate numbers that are 10 greater than a given number, 10 less, 20 greater, and so on (15).

A Wider Role

Estimation is not a rigidly defined skill; flexibility and timing seem to be hallmarks of teachers who use it well in instruction. Consequently, research studies steer away from rigid guidelines for the use of estimation in the classroom. The tenor of their recommendations is that estimation should not be treated as a separate topic but should be woven into all parts of the mathematics curriculum. *Mental arithmetic*, here referring to computational techniques specifically designed for use without pencil and paper, lends itself more to systematic presentation than does estimation, and that is reflected somewhat in the research recommendations. For both skills, however, one message is clear: they should play a wider role in classroom instruction.

A Wider Role: Estimation

In the interpretation of the 1975 results from the National Assessment of Educational Progress (NAEP), the reviewers stated that "too many students simply learn estimation as one more abstract computational device in which for some reason or other an arbitrary number of digits is replaced by zeros" (3). That is unfortunate, because estimating skills can enrich counting, measurement, and problem solving experiences. Developed early, they can give teachers a diagnostic boost as they search for the causes of systematic errors. Used more widely, estimation skills can even provide creative approaches to computation. For example, in the introduction of algorithms, such as that for the addition of fractions, a teacher could ask: "How large will 1/2 + 1/4 be? Larger than 1/2? Yes, so we can't find the answer just by adding numerators and adding denominators (1 + 1 and 2 + 4). This would equal 2/6, which equals 1/3, which is smaller than 1/2." The more frequently teachers encourage their students to think about the reasonableness of answers, the more prepared those students will be to reject answers that make no sense in the context of particular problems.

More discussion of estimation in *problem solving* appears in Johnson (6) and O'Daffer (9) and in the *Research Within Reach* bulletin "Mathematical Problem Solving: Not Just a Matter of Words." Estimation in *measurement* gets extensive treatment in Bright's article (2). Examples for enriching *counting* and *computational* work with estimation appear in the articles by Johnson (6), O'Daffer (9), and Trafton (15). Trafton presents a set of teacher guidelines for work in estimation.

A few cautionary notes do emerge from the recommendations of researchers:

- Children should be encouraged to develop their own ways of deciding when an answer is reasonable. Later, teachers may help children refine their techniques or show them more standard procedures, like rounding to the nearest ten (1). One effective procedure is to give students problems for which an answer is provided and encourage them to discuss whether it is reasonable, without calculating exactly (6).

- At first, tolerate wild guesses, but gradually guide the children to more systematic estimating techniques. There is never any *best estimate*, so be accepting of roughly cut approximations. Too frequently, children get the idea that an estimate is good *only* if it's the exact answer. Some discussion of how close an estimate should be in various situations can help students determine what a "good" estimate is.

- Don't require children to check out the correctness of their estimates by following through with complete calculations. Discussions about the validity of estimates should grow from a consideration of the estimating procedures themselves (3).

A Wider Role: Mental Arithmetic

Having surveyed the research literature on mental arithmetic use in the classroom, one writer concluded: "The data are consistent and fairly conclusive that mental computation instruction produces good results in general arithmetic growth" (16).

In particular, one sixth-grade study (4) produced convincing evidence that ten minutes a day of training in mental computation, over a period of several months, improved both mental computation *and* problem solving skills *for fast and slow students*. In another study, similar training at the fifth-grade level (ten minutes a day for twenty school days) resulted in gains on a problem solving retention test (12).

What kinds of training in mental arithmetic should teachers consider? Two skills receiving attention and emphasis are *renaming mentally* and *compensation* (13). Two examples of renaming mentally follow—the first problem is 617 + 352 = ?

"Let's see. 617 is 600 + 17 and
              352 is 350 + 2.
 Now,   600 + 350 = 950 and
              17 +   2 = 19.
 So,   950 + 19 = 969."

"I know
$$6 \times 72 = ?$$
72 is 70 + 2, that
$6 \times 70 = 420$, and that
$6 \times 2 = 12$.
So, $6 \times 72 = 420 + 12 = 432$."

Next follow two examples of *compensation*:

$$212 - 89 = ?$$

"If I add 11 to 89 that will make 100, which is easy to work with. But I'll have to add 11 to 212, also. Then I get 223 − 100, which is 123, so
$$212 - 89 = 123."$$

$$6 \times 99 = ?$$

"That's pretty close to 6×100, which is 600. In fact, it's just one 6 less than 600. So, $6 \times 99 = 594$."

Mental arithmetic skills like renaming mentally and compensation take root most firmly if they have an individual flavor and reflect individual student choices. Rather than enforcing a rigid approach in the applications of either skill, teachers should encourage each student to develop his or her own manner of applying them to problems.

Conclusion

The research literature is clear on several points: estimation and mental arithmetic skills *can* be taught; they can be taught to *all* students; they can and should be developed *early* and *regularly*. It is also important to note that there is evidence that regular exposure and practice has a positive effect on student *attitudes* toward mathematics (12).

With the advent of increased classroom use of hand-held calculators, estimating skills take on new importance, because they offer children more control over the machines ("Did you press all the right keys? Can you tell from the size of the answer shown?"). In fact, an alliance of mental arithmetic, estimation, and calculator use can bring both strength and flexibility to the development of mathematical problem solving skills.

The most recent NAEP results provide further evidence that elementary school children need stronger and more flexible estimating skills. The results for the following problem are striking (8):

> A man has 1,130 baseballs to pack in boxes which hold 24 baseballs each. How many baseballs will be left over after the man has filled as many as he can?

More than 25 percent of the thirteen year olds who worked on this problem chose answers that were not even whole numbers! Clearly, the need for estimation training outlined in this bulletin is a critical need.

The real world, as any adult knows, is not always tolerant of paper-and-pencil dependency in mathematics. In fact, people use estimation and mental arithmetic far more frequently than they use paper-and-pencil computation. There is no reason, therefore, for those who create the child's world in school to be any more tolerant of paper-and-pencil dependency.

References

1. Ashlock, R.B. *Error Patterns in Computation.* Columbus, Ohio: Charles E. Merrill, 1972.
* 2. Bright, George W. "Estimation as Part of Learning to Measure." In Doyal Nelson and Robert E. Reys (eds.), *Measurement in School Mathematics,* 1976 NCTM Yearbook. Reston, Virginia: National Council of Teachers of Mathematics (NCTM), 1976.
* 3. Carpenter, Thomas P., et al. "Notes from National Assessment: Estimation." *The Arithmetic Teacher,* 23 (April 1976): 297-302.
4. Flournoy, Mary Frances. "The Effectiveness of Instruction in Mental Arithmetic." *Elementary School Journal,* 55 (November 1954): 148-153.
5. Ginsburg, Herbert, "Young Children's Informal Knowledge of Mathematics." *The Journal of Children's Mathematical Behavior,* 1 (Summer 1975): 63-156.

References (continued)

6. Johnson, David C. "Teaching Estimation and Reasonableness of Results." *The Arithmetic Teacher,* 27 (September 1979): 34-36.

7. Kramer, Klaas. "Adding and Subtracting Without Pencil and Paper." In Klaas Kramer (ed.), *Problems in the Teaching of Elementary School Mathematics.* Boston, Massachusetts: Allyn and Bacon, 1970.

8. National Assessment of Educational Progress. *The Second Assessment of Mathematics, 1977-78: Released Exercise Set.* Denver, Colorado: Education Commission of the States, 1979.

*9. O'Daffer, Phares. "A Case and Techniques for Estimation: Estimation Experiences in Elementary School Mathematics—Essential, Not Extra!" *The Arithmetic Teacher,* 26 (February 1979): 46-53.

*10. Payne, Joseph, and Edward Rathmell. "Number and Numeration." In Joseph N. Payne (ed.), *Mathematics Learning in Early Childhood,* 37th NCTM Yearbook. Reston, Virginia: NCTM, 1975.

*11. Rathmell, Edward C. "Using Thinking Strategies to Teach Basic Facts." In M. Suydam and R. Reys (eds.), *Developing Computational Skills,* 1978 NCTM Yearbook. Reston, Virginia: NCTM, 1978.

12. Schall, William E. "Comparing Mental Arithmetic Modes of Presentation in Elementary School Mathematics." *School Science and Mathematics,* 73 (May 1973): 359-367.

13. Suydam, Marilyn N., and Dessart, Donald J. *Classroom Ideas from Research on Computational Skills.* Reston, Virginia: NCTM, 1976.

14. Thornton, Carol A. "Emphasizing Thinking Strategies in Basic Fact Instruction." *Journal for Research in Mathematics Education,* 9 (May 1978): 214-227.

*15. Trafton, Paul R. "Estimation and Mental Arithmetic: Important Components of Computation." In M. Suydam and R. Reys (eds.), *Developing Computational Skills,* 1978 NCTM Yearbook. Reston, Virginia: NCTM, 1978.

16. Zepp, Raymond. "Algorithms and Mental Computation." In Marilyn N. Suydam and Alan R. Osborne (eds.), *Algorithmic Learning.* Columbus, Ohio: The ERIC Clearinghouse for Science, Mathematics and Environmental Education, 1976. (ED 113 152).

*These are the readings we believe would be especially interesting to and useful for teachers.

Research Within Reach
Elementary School Mathematics

Calculators in the Classroom

Question: Are my second-grade students too young to benefit from the use of handheld calculators during classroom instruction?

Since most of the research is recent, the long-term effects of calculators in the classroom are not yet clear. There is, however, substantial evidence that points to the calculator as an exciting and useful tool for teaching problem solving, estimation, reasonableness—indeed, for developing concepts across the entire mathematics curriculum.

In the past five years nearly 100 research studies have compared noncalculator groups with groups that used calculators on some regular basis in the classroom. Of those studies, very few showed an advantage in mathematics achievement for the noncalculator groups. The vast majority of studies revealed either no significant difference or an advantage for the calculator groups (16).

The most recent National Assessment of Educational Progress (NAEP) provides some indication that the use of calculators, in and of itself, carries no guarantee of success in either computational exercises or problem solving (11). A number of students tested by NAEP worked a series of problems with calculators. A comparable group worked the same problems without calculators. On exercises involving straight computation, results were generally mixed. For example, computations that involved decimals proved to be more troublesome for the users of calculators than for the non-users. On the other hand, calculator users had an easier time with straight division problems.

In the category of word problems, results again were mixed. On problems involving division with whole-number answers, calculator users had, in general, a higher success rate than nonusers. However, on problems like the one below, where there is a division remainder, non-users had the better record.

112

A man has 1,310 baseballs to pack in boxes which hold 24 baseballs each. How many baseballs will be left over after the man has filled as many as he can?

Research is showing the calculator to be a powerful teaching and learning tool. Clearly, however, students will need training and practice in interpreting calculator answers in the light of different problem contexts.

Teachers' and parents' concern about the possible negative effects of the use of calculators will not totally vanish, of course, and the need for further research is universally acknowledged. Furthermore, like any instrument of potential good, the calculator is not immune to misuse. The purpose of this bulletin is not to provide the last word but rather to provide suggestions for creative and profitable calculator use by elementary school teachers.

Calculators in the Primary Grades

Mathematical learning for young children is tightly tied to sense perceptions and concrete experiences. With this in mind, Ginsburg (5) has stressed the need for young children to see regularity in mathematics. Calculators can provide a number of visual examples of regularity. First graders learn to use calculators very quickly, "usually within the first hour" (3). After that, they seem to become alert to mathematical regularity, even on the very basic level of repeated counting; while they press the calculator repeat button (on many, the = button), they count out loud as the calculator displays the numbers. When they begin working with mathematical operations, it is important that they appreciate, for example, that all multiples of 5 end either in 0 or 5, and that counting by .1 and counting by 1 result in similar patterns. Through all these experiences, the display screen of a calculator can assist teachers in pointing out mathematical regularity.

Several studies have reported the enthusiasm with which young children take to calculators. There is evidence of an increase in confidence, improved attitudes toward mathematics, and greater persistence in problem solving. Furthermore, far from becoming calculator-dependent, young children tend, as the machine's novelty wears thin, to use calculators less often in exercises that are familiar. In comparison with peers who are less experienced with calculators, they tend to make wiser decisions concerning calculator use. (See references 9, 10, 17, and 18.)

Many suggestions for using calculators with primary level children are presented in research we cite (2, 4, 7, 14). We give one example here, from Beardslee (2), for reinforcing children's understanding of place value:

Game: Replace by Zero (2 players)

Materials Needed: 2 calculators, score sheet

Procedures: Player A selects a three-digit number and directs player B to enter it into the calculator. Then player A directs player B to replace one of the digits by 0 without affecting any of the other digits. (For example, enter 273. Replace the 7 by 0. The student should subtract 70, leaving 203 in the display.) Players take turns.

A recent study (18) confirmed that such calculator place-value exercises can be a profitable undertaking as early as the second grade. The researchers also found that calculators enable very young children to focus on estimation strategies and to solve multiplication and division word problems before they become proficient in the computational procedures of multiplication and division.

Another form of research is the survey, and recent surveys sketch an interesting picture. Suydam (16) reports that nearly 80 percent of all children, elementary through secondary, have access to calculators. Eight-five percent of teachers in Reys' survey (13) said calculators should be available to students in school.

Finally, a 1977 survey of more than 1300 teachers led to this conclusion: "In the first grade, calculators were used most frequently for drill. . . . Above first grade the most frequent usage was for checking" (6). Drill and checking are important components of any elementary school curriculum. However, teachers who concentrate their calculator use on these two aspects miss some very fertile ground for teaching.

As we have seen, the research shows that calculators can be used, should be used, and *will* be used. The challenge lies in using them well.

Using Calculators Well

We argue in the *Research Within Reach* bulletin, "Mathematical Problem Solving: Not Just a Matter of Words," that problem solving should be an integral part of the entire elementary school mathematics curriculum and not appear only in the guise of approaches to working out textbook word problems.

Successful problem solvers work comfortably and productively with estimating, retracing solution steps, and questioning the reasonableness of answers. In one recent study (17), researchers looked for the use of those skills (and others) in a sixth grade, calculator vs. noncalculator, problem solving setting. Both groups of students received regular instruction and practice in problem solving strategies like estimating, retracing, and checking reasonableness, but the calculator group applied those skills much more frequently than did the noncalculator group.

- Computational skills often dominate a mathematics program. The use of calculators can ease the burden for students with computational weaknesses and help to provide a mathematics program with more richness and depth. In particular, calculators hold tremendous potential for helping low-achieving students as they concentrate on the problem solving process. Several research studies have underscored this potential. For example, in one study of sixth graders, students were given a set of problems and were then presented with problem solving questions that tested their ability to determine which computational procedures were necessary to solve the problems. The researchers found that the use of calculators resulted in average and low ability students correctly answering a greater number of questions than comparable students who did not use calculators (8).

In *Overview and Analysis of School Mathematics: Grades K-12* (12), the National Advisory Committee on Mathematical Education (NACOME) takes note of the potential that calculators have for improving *all* students' mathematics learning, remarking that calculators "allow students to feel the power of mathematics and free time for teachers to concentrate on the conceptual aspects of the subject which are of fundamental importance" (p. 43).

- As children become more polished in their estimating skills, calculators can be used creatively to reinforce those skills (see the *Research Within Reach* bulletin "Estimation and Mental Arithmetic" for an account of the importance of estimating skills). The following exercise from Reys (14) is just one example:

    Estimate the product. Circle the answer that is closest to your estimate. Use your calculator to find the actual product, then subtract for the difference.

    $$3 \times 38 = \underline{} \quad \begin{array}{c} 130 \\ 120 \\ 90 \\ 60 \end{array}$$

    $$\underline{} - \underline{} = \underline{}$$
    Estimate   Actual   Difference
             Product

Conclusion

Introducing calculators into the curriculum requires planning—for purchase, maintenance, effective use, and so on. Guidelines for buying and maintaining calculators appear in a number of sources, and we have listed two (3 and 4) in our references section.

It is clear that the use of calculators in the classroom has to reflect teachers' individual styles. Furthermore, despite the wealth of recent literature on calculator use, the role the calculator can play in concept development has yet to be finally determined. As a result, and until that role becomes clearer, teachers who are interested in using calculators in their classrooms will have to proceed, to some extent, by trial and error.

Several things *are* clear right now. Calculators are widely available, children can learn with them, and they *want* to learn with them. The major question facing educators is not, Should we use calculators? but, Where and how can calculators best be used to help children learn mathematics?

References

1. Aidala, Gregory. "Calculators: Their Use in the Classroom." *School Science and Mathematics*, 73 (April 1978): 307-312.
* 2. Beardslee, Edward C. "Teaching Computational Skills with a Calculator." In M. Suydam and R. Reys (eds.), *Developing Computational Skills*, 1978 NCTM Yearbook. Reston, Virginia: National Council of Teachers of Mathematics (NCTM), 1978.
3. Bell, Max S. "Calculators in Elementary Schools? Some Tentative Guidelines and Questions Based on Classroom Experience." *The Arithmetic Teacher*, Special Issue: Mini Calculators, 23 (November 1976): 502-510.
4. Caravella, Joseph R. *Minicalculators in the Classroom*. Washington, D.C.: National Education Association, 1977.
5. Ginsburg, Herbert. "Young Children's Informal Knowledge of Mathematics." *The Journal of Children's Mathematical Behavior*, 1 (Summer 1975): 63-156.
6. Graeber, Anna O.; Eui-Do Rim; and Nancy J. Unks. *A Survey of Classroom Practices in Mathematics: Reports of First, Third, Fifth and Seventh Grade Teachers in Delaware, New Jersey, and Pennsylvania*. Philadelphia, Pennsylvania: Research for Better Schools, Inc., 1977.
7. Immerzeel, George, and Earl Ockenga. *Calculator Activities for the Classroom*, Book 1. Palo Alto, California: Creative Publications, Inc., 1977.
8. Kasnic, Michael J. "The Effect of Using Handheld Calculators on Mathematics Problem-Solving Ability among Sixth Grade Students." Ph.D. Dissertation, Oklahoma State University, 1977.
9. Mason, Marguerite. "The Effects of a Year Long Program of Calculator-Aided Mathematics Instruction on the Computational Strategies of Elementary School Children." Paper presented at the annual meeting of the National Council of Teachers of Mathematics, Boston, April 1979.
10. Moser, James M. "The Effect of Calculator-Supplemented Instruction upon the Arithmetic Achievement of Second and Third Graders." Paper presented at the annual meeting of the National Council of Teachers of Mathematics, Boston, April 1979. (Available from Wisconsin Research & Development Center for Individualized Schooling, Madison, Wisconsin).
11. National Assessment of Educational Progress. *The Second Assessment of Mathematics, 1977-78: Released Exercise Set*. Denver, Colorado: Education Commission of the States, 1979.
12. *Overview and Analysis of School Mathematics: Grades K-12*. Washington, D.C.: Conference Board of the Mathematical Sciences, 1975.
13. Reys, Robert E., et al. "Hand Calculators—What's Happening in Schools Today?" *Arithmetic Teacher*, February 1980, in press.
14. Reys, Robert E., et al. *Keystrokes: Calculator Activities for Young Students. Multiplication and Division*. Palo Alto, California: Creative Publications, Inc., 1979.
15. Suydam, Marilyn N. *Investigations with Calculators: Abstracts and Critical Analyses of Research*. Columbus, Ohio: Calculator Information Center, 1979.
*16. Suydam, Marilyn N. *The Use of Calculators in Pre-College Education: A State-of-the-Art Review*. Columbus, Ohio: Calculator Information Center, 1979.
17. Wheatley, Charlotte L. "The Effect of Calculator Use on the Problem Solving Strategies of Elementary School Pupils." Paper presented at the annual meeting of the National Council of Teachers of Mathematics, Boston, April 1979.
*18. Wheatley, Grayson, et al. "Calculators in Elementary Schools." *Arithmetic Teacher*, 27 (September 1979): 18-21.

*These are the readings we believe would be especially interesting to and useful for teachers.

Research Within Reach
Elementary School Mathematics

Sequence with Substance:
The Elementary School Mathematics Curriculum

Question: Why are we expected to make the children do so much so soon—for example, beginning subtraction in the first few weeks of first grade, before they are used to addition?

Within classroom walls, when learning is on the line, decisions lie in teachers' hands. And yet many of the concepts and skills they are expected to teach, and the order in which they are expected to teach them, are related to a string of research and development decisions in which teachers have had little, if any, part. As a result, many teachers are curious about the ways in which curriculum decisions are made, and they occasionally question the wisdom in some of the decisions.

How much geometry should be taught to elementary school children, and when? Should fractions be taught before decimals? When should children begin work on word problems? These, along with the question which introduces this bulletin, are but a few of the curriculum questions teachers have asked. The following pages will profile the curriculum research in elementary school mathematics, with an eye toward answering the questions that can be answered and alerting readers to important areas of current investigation.

Curriculum questions, by their very nature, tend to deal in a broad way with the whats, wheres, whens, and whys of school teaching and learning. The search for answers requires a range of concerns just as broad. For example, since mathematics lends itself to the logical arrangement of concepts, *the logical sequencing of mathematical concepts in the elementary school curriculum demands attention.*

An equally important concern is the *coordination of the curriculum with child development,* since the logical structure of mathematical concepts and the cognitive processes a child goes through in acquiring the concepts may differ. For example, $½+⅓$ is taught before $½×⅓$, even though the algorithmic process for multiplying

116

fractions is simpler than that for addition. However, when an individual attempts to derive meaning from the steps of the two algorithms, it is multiplication that demands a more sophisticated level of mathematical understanding.

Finally, there is another source of curriculum concern. Since society views school curriculum in light of its usefulness in meeting the needs of every citizen, educators must also *consider how well the mathematics curriculum paves the way for learning skills which endure* and which continue to bring rewards in the world outside the classroom. This area does not have a strong research base, although a number of researchers are looking at how well experience with real-life applications improves problem solving ability. Their conclusions may affect future curriculum decisions.

Approaches to Sequencing the Mathematics Curriculum

Much of the research in sequencing the mathematics curriculum takes direction from two distinct perspectives. One perspective grows out of the theory of learning hierarchies and is related to the logical arrangement of concepts mentioned above. According to this theory, illustrated in Example 1 below, all higher-order skills and knowledge develop from lower-order skills and knowledge, and a careful teaching sequence of the subconcepts in a hierarchy can result in smooth transfer of learning from one level to the next, until the higher-order concept is learned.

The second perspective derives mostly from the child development research of Piaget (9) and Bruner (1), in particular the conviction that children can understand mathematical concepts in an intuitive way well before they grasp them in an abstract, more general manner. This perspective leads some researchers to try to determine effective sequences of instruction that will guide children from concrete, intuitive introductions to a particular mathematical concept (the so-called enactive level), to an understanding of pictorial representations of the same concept (the iconic level), and finally to a grasp of the concept on an abstract (symbolic) level.

As an illustration of the differences between the two perspectives, consider the following two sequences aimed at the same learning goal, adding fractions with like denominators.

Example 1. Using a procedure developed by Gagné, the researchers in one study (10) generated a learning hierarchy of eleven steps for the task of adding fractions with like denominators:

Subtasks	Examples
1. Adding with two fractions having like denominators where the sum requires no renaming.	$2/9 + 3/9 = ?$
2. Adding with three fractions having like denominators where the sum requires no renaming.	$1/10 + 3/10 + 5/10 = ?$
3. Adding a whole number and a rational number named by a fraction.	$2 + 1/2 = ?$
4. Adding with two mixed numerals having like denominators.	$3\ 1/7 + 2\ 3/7 = ?$
5. Finding equivalent fractions by dividing both numerator and denominator by the same number.	$6/9 = ?/3$
6. Adding with two fractions with like denominators where the sum requires rewriting in simplest form.	$3/16 + 5/16 = ?$
7. Adding with two mixed numerals with like denominators where the sum of the fractional parts requires rewriting in simplest form.	$4\ 1/9 + 5\ 2/9 = ?$
8. Adding with three mixed numerals where all de-	$3\ 1/6 + 4\ 1/6 + 2\ 2/6 = ?$

nominators are alike and the sum of the fractional parts requires rewriting in simplest form.

9. Changing names from an improper fraction to a mixed numeral. $12/8 = 8/8 + ?/8 = ?$

10. Adding with two mixed numerals where the denominators involved are alike and the sum of the fractional parts is greater than 1. $3\ 5/8 + 2\ 7/8 = ?$

11. Adding with mixed numerals where the denominators involved are alike and the sum of the fractional parts is greater than one. $7\ 2/5 + 4\ 1/5 + 1\ 4/5 = ?$

Example 2. Here are highlights of a teaching sequence, from initial fraction concepts through the addition of like fractions, which reflects recent child development research. Ellerbruch and Payne's article (6) should be read for details, as the sequence is extended to include fractions with unlike denominators.

1. Use a *single* mathematical model for the introduction of fractions, and stick with it—for example, rectangular strips of paper, initially in some familiar form, say, candy bar wrappers.

2. Using two candy bar wrappers of different sizes, get the children used to the size of the *unit* ("Which candy bar would you choose?" "Why?").

3. Using two candy bar wrappers of the same size (or substitute rectangular strips), cut one into 3 equal pieces, the other into 3 unequal pieces. Help the children to recognize that a "fair cut" is made only if the pieces in a divided rectangle are the same size.

4. Using two rectangular pieces of the same size, split each into a different number of equal-sized pieces. Help the students to observe that the size of a piece is determined by the *number* of pieces: the more pieces, the smaller the piece.

5. After a period of time in which the students compare different fractions by folding same-sized rectangular strips, begin to teach the oral names of fractions and provide practice for the students in pronouncing and using those names.

6. Move to the visual level of understanding by repeating steps 1-4 with blackboard or overhead projector representations of fractions.

7. Begin to guide the students to symbolic representation. To quote the researchers: "Display a strip to show three-fourths. Ask the usual four questions: 'What is the unit? Are the pieces of equal size? How many pieces am I holding up (fold back one)? How many pieces in the unit?' Then ask for the oral name when you display three-fourths: 'What is the name for this amount?' Write '3 fourths' on the chalkboard. Then show that a short form of writing this is 3/4." Repeat this procedure for other fractions.

8. Move, finally, to addition of fractions with like denominators, first in concrete fashion by having the students pair off, one in each pair holding two fourths in rectangular strips, the other holding one fourth. Ask questions like "How big is each piece?" "How much are you holding up together? Name it with a fraction."

Extend their understanding to a symbolic level by translating what they've just done into the following 4-step procedure for adding like fractions: (1) Decide if the pieces are the same size. Are the bottom numbers the same? If not, write "can't solve." (2) Add the top numbers and write the answer. (3) Draw a line underneath. (4) Write the bottom number underneath the line.

Curriculum Research Findings

We now list some conclusions and recommendations growing out of research on the elementary school mathematics curriculum. Several of these have direct implications for classroom teaching; others, as we've noted, will have to filter through development programs and textbook design before they can affect classroom activities. The first group of findings concerns sequencing:

- Teaching subtraction facts (13−4=9) along with their related addition facts (4+9=13) leads to better understanding, provided there is a consistent emphasis on the relationship between the two operations (14).

- Initial learning at the concrete level aids in the learning of addition, subtraction, multiplication, fractions, and problem solving (15). Whether the concrete-pictorial-symbolic sequence is necessary for *every* mathematical topic is not clear, nor is it clear whether the sequence can be shortened for some concepts or at particular learner age levels.

  What *is* clear, however, is that at all elementary school levels the use of purely symbolic treatments of mathematical topics is not as effective as the use of sequences in which manipulative materials are used (15).

- Phillips and Kane (10) compared the teaching effectiveness of the eleven-step learning hierarchy outlined above with the effects of six other sequences of the same eleven subtasks, all of which were introduced with programmed instruction. Their study indicated that learning retention is vulnerable to sequencing, and that the best retention from the seven sequences they tested resulted from the several sequences generated according to learning hierarchies theory.

  This study suggests that mathematics curriculum designers should give careful consideration to the ordering of prerequisites, even on a finely graded subconcept level.

- Sequencing for the learning of algorithms is as important as sequencing for the learning of mathematical concepts, such as place value and the addition of fractions. With a blending of hierarchical and representational (that is, concrete, pictorial, and symbolic) considerations, Hazekamp (7) presents a sequence for teaching the conventional two-digit multiplication algorithm. He also lists a hierarchy of steps derived from his own research. Some of these steps are:

  1. Knowledge of basic multiplication facts.
  2. The structure of the base-ten system, and renaming skills.
  3. Multiplying multiples of 10 and multiples of 100 by ones.
  4. ....

  Within each step Hazekamp presents a number of suggestions for mastery on concrete, pictorial, and symbolic levels. For example, in step 3, base-ten blocks (single small cubes are the ones blocks, strips of 10 ones blocks together are the tens blocks, and so on) are used to illustrate the connection between $3 \times 4 = 12$ and $30 \times 4 = 120$.

- There is more than one way to approach most mathematics topics, including sequencing the teaching of fractions. A clear recommendation from research, however, is that computation with fractions should not begin until children have a firm grasp on fraction concepts—for example, the idea of a unit, the part-whole relationship, and the relative size of fractional parts (4).

- Children can do some learning of fractions in primary grades, even the first grade. In particular, Ellerbruch and Payne (6) report success in primary grades with steps 1-5 of their sequence outlined above, that is, the concrete and oral language steps, with occasional success at step 6, the visual representation of fractions. The instruction in initial concepts and oral language takes about 7-10 days at the primary level.

- When oral names of fractions are introduced *before* written symbols, there is a marked reduction in the number of reversal errors later on—for example, 3/4 is less often mistakenly written as 4/3 (6).

- There is no strong research evidence favoring the teaching of fractions before decimals, or decimals before fractions (14). This may change, however, as the use of hand calculators in developing estimation and problem solving skills increases, resulting in an earlier introduction of decimals into the curriculum.

Other Areas of Curriculum Investigation

Education researchers have an ongoing interest in mapping curriculum sequences that are conducive to mathematics learning. Yet not all pieces of the map fall easily into place.

For example, the selection and sequence of geometry concepts in elementary school mathematics has attracted research attention, but few specifics have been resolved. A clear and firm conclusion, however, is that geometry *belongs* in the elementary school curriculum, even at the early primary level. Indeed, a number of studies have indicated that elementary school children can learn a variety of geometry topics, including concepts of topology (such as shapes, inside and outside, line connectedness), motion geometry, coordinate geometry, and simple construction (5; also see 12).

Research also points to the value of nurturing the learning of geometry with regular use of activities, discovery, guessing, and problem solving (2).

In the area of problem solving, there is evidence that once children master a strategy appropriate for solving one type of word problem, they can easily get so locked into that strategy that they will apply it in other problem situations when it is inappropriate. Researchers, aware that instruction can do much to offset these traps, have begun to look for helpful sequences of both word problem types and problem solving strategies (8). Until research provides more clarity and direction, teachers can do some mixing of problems as they present them to their students, making certain that a single strategy will not work for each group of problems.

The Quality of Learning within the Curriculum

Sequencing concepts in the mathematics curriculum is one thing; helping children to tie those concepts together in rich and enduring fashion is another. Here are some research suggestions for making the curriculum more cohesive:

- Both the retention of basic facts and the ease with which those facts are used to learn harder facts will improve if children are taught thinking strategies in conjunction with both concrete activities and drill (17). For example, Rathmell (11) reports three effective thinking strategies for addition which can, in combined use, make learning the 100 basic addition facts ($0+0=0$ through $9+9=18$) easier:

    1. The "counting on" strategy ("$2+9=?$ Counting on from 9, I get 9 . . . 10, 11").
    2. The "one more or one less" strategy ("$5+6=?$ I know $5+5=10$, so $5+6$ is one more, so $5+6=11$").
    3. The "compensation" strategy, by which the two numbers being added are changed in compensating fashion to yield a more familiar fact ("$9+6=?$ 9 and 1 is 10, 10 and 5 more is 15, so $9+6=15$").

    If there is careful modelling of the thinking strategies by the teacher and ample practice for the students, the time required at the drill/memorization stage diminishes.

- The teaching of thinking strategies is only one way to enrich the mathematics curriculum. Work with estimation, pattern finding, and problem solving should be woven into the curriculum *as early as possible* in the child's school experience, not as separate topics but as learning techniques to be applied whenever situations invite their use (18). A number of relevant suggestions are in the *Research Within Reach* bulletins "Estimation and Mental Arithmetic" and "Mathematical Problem Solving: Not Just a Matter of Words."

- Make frequent efforts to illustrate the continuity of the mathematics curriculum (for example, regular references to the relationships between addition and subtraction, between the addition of fractions and the addition of whole numbers, and between fractions represented by sections of a circle and fractions represented by sections of a ruler). There is evidence that teacher efforts to point out continuity have a beneficial effect on student attitude toward mathematics (3).

- Finally, check the *Research Within Reach* bulletin "Meaning in Elementary School Mathematics" and weigh the considerable evidence there which connects superior student retention and transfer of learning, on the one hand, with the teaching of mathematics for understanding and meaning, on the other. For example, teachers should take pains that children understand the mathematical reason for bringing a one to the tens column in the example below, not just satisfy themselves that the children perform the task correctly:

$$\begin{array}{r} 34 \\ +28 \\ \hline 62 \end{array}$$

Good retention brings a sense of unity to the learning of mathematics; effective transfer implies a smooth flow from one concept to another. Both are essential ingredients for a cohesive curriculum.

Further Recommendations

Educational research is only one force affecting curriculum decisions. Other considerations, political, philosophical, and cultural, are bound to come into play in such an important area.

In the face of such considerations and decisions, a number of organizations with national stature have made curriculum recommendations. We end this bulletin with the basic skills list developed by the National Council of Supervisors of Mathematics (13). According to NCSM, the goal of the mathematics curriculum should be to ensure that each student is able to:

1. Solve problems.
2. Apply mathematics to everyday situations.
3. Be alert to the reasonableness of results.
4. Estimate and approximate.
5. Compute with appropriate skills.
6. Use geometry.
7. Measure.
8. Read, interpret, and construct tables, charts, and graphs.
9. Use mathematics to predict.
10. Understand the role of computers.

References

1. Bruner, Jerome. *Toward a Theory of Instruction.* New York: W.W. Norton, 1966.

* 2. Callahan, Leroy G., and Vincent J. Glennon. *Elementary School Mathematics: A Guide to Current Research.* 4th edition. Washington, D.C.: Association for Supervision and Curriculum Development, 1975.

3. Campbell, N. Jo, and Harold L. Schoen. "Relationships between Selected Teacher Behaviors of Prealgebra Teachers and Selected Characteristics of Their Students." *Journal for Research in Mathematics Education,* 8 (November 1977): 369-375.

4. Carpenter, Thomas P., et al. "Notes from National Assessment: Addition and Multiplication with Fractions." *The Arithmetic Teacher,* 23 (February 1976): 137-144.

* 5. Downes, John P., Rosalie S. Jensen, and Hiram D. Johnson. *76 Questions: A Synthesis of the Research on Teaching and Learning Mathematics.* 1977, 119 pp. ED 162 896.

6. Ellerbruch, Larry W., and Joseph N. Payne. "A Teaching Sequence from Initial Fractions." In Marilyn N. Suydam and Robert E. Reys (eds.), *Developing Computational Skills,* 1978 NCTM Yearboook. Reston, Virginia: National Council of Teachers of Mathematics (NCTM), 1978.

7. Hazekamp, Donald W. "Teaching Multiplication and Division Algorithms." In M. Suydam and R. Reys (eds.), *Developing Computational Skills,* 1978 NCTM Yearbook. Reston, Virginia: NCTM, 1978.

8. Mathematics Resource Project. *Didactics and Mathematics.* Palo Alto, California: Creative Publications, Inc., 1978.

* 9. Payne, Joseph N. (Ed.) *Mathematics Learning in Early Childhood,* 37th NCTM Yearbook. Reston, Virginia: NCTM, 1975.

10. Phillips, E. Ray, and Robert B. Kane. "Validating Learning Hierarchies for Sequencing Mathematical Tasks in Elementary School Mathematics." *Journal for Research in Mathematics Education,* 4 (May 1973): 141-151.

*11. Rathmell, Edward C. "Using Teaching Strategies to Teach the Basic Facts." In M. Suydam and R.

References (continued)

Reys (eds.), *Developing Computational Skills*, 1978 NCTM Yearbook. Reston, Virginia: NCTM, 1978.

12. Robinson, G. Edith. "Geometry." In J.N. Payne (ed.), *Mathematics Learning in Early Childhood*, 37th NCTM Yearbook. Reston, Virginia: NCTM, 1975.

13. Suydam, Marilyn N. "The Case for a Comprehensive Mathematics Curriculum." *The Arithmetic Teacher*, 26 (February 1979): 10-13.

*14. Suydam, Marilyn N., and Donald J. Dessart. *Classroom Ideas from Research on Computational Skills*. Reston, Virginia: NCTM, 1976.

15. Suydam, Marilyn N., and Jon L. Higgins. *Activity-based Learning in Elementary School Mathematics: Recommendations from Research*. Reston, Virginia: NCTM, 1977.

16. Suydam, Marilyn N., and J. Fred Weaver. *Using Research: A Key to Elementary School Mathematics*. Reston, Virginia: NCTM, 1975.

17. Thornton, Carol A. "Emphasizing Thinking Strategies in Basic Fact Instruction. *Journal for Research in Mathematics Education*, 9 (May 1978): 214-227.

18. Trafton, Paul R. "The Curriculum." In J.N. Payne (ed.), *Mathematics Learning in Early Childhood*, 37th NCTM Yearbook. Reston, Virginia: NCTM, 1975.

---

*The references with asterisks are readings we believe would be especially interesting to and useful for teachers.

Research Within Reach
Elementary School Mathematics

The Teacher and the Textbook

Question: Our current text stresses drill, and I am afraid my students will get shortchanged in understanding. In which directions, and for which topics, should the text be supplemented?

It is important to keep textbooks in perspective. By themselves, they have little power to motivate students, to raise self-concepts, or to cultivate problem-solving instincts and skills. A good text can contribute to such growth, but the power to make it happen must come from the teacher.

Answers to teacher questions about textbooks depend upon the texts and the students involved. You must be the final judge in deciding if and when your text is missing the mark on a particular topic. Recent research, however, has drawn some conclusions and made some recommendations concerning the use of textbooks in elementary school mathematics, making it possible to approach questions on textbook use with a more analytic eye. This bulletin looks at current textbook design and usage, at some textbook pitfalls and weaknesses that can be anticipated, and at sources of advice for teachers in the wise use of texts.

Textbook Use

Research surveys and analyses of mathematics textbooks and their use reveal some expected findings, as well as a few surprises. In their survey of the research in the twenty-year span from 1955 to 1975, Suydam and Osborne found that a single text is used in most classrooms and that many teachers use no instructional materials except the textbook and the chalkboard (15).

The major mathematics texts in use throughout the country are similar in several ways:

> Low-level cognitive processes—knowledge and comprehension—are used far more frequently than high-level processes—for example, problem-solving processes (15).

An emphasis on computational skills is apparent (5, 15).

There is considerable agreement on grade placement, sequence, and presentation of basic topics (15).

Of the 32 most widely used mathematics materials, including textbooks, 29 are based on "conventional wisdom" rather than research and development (4).

Despite the similarities in broad goals and areas of emphasis (for example, whole number computational skills), some significant differences do exist among major textbooks. In particular, there is a wide difference in the total number of concepts treated and the amount of space devoted to various topics (15). One recent analysis of three widely used fourth-grade textbooks—Addison-Wesley, Houghton-Mifflin, and Scott-Foresman—provided a detailed account of the variance. The researchers first compiled a list of all the mathematics topics covered in these texts. A comparison of the texts disclosed that more than *half* of the topics on this list are covered in only one of the three texts (8). For example, ordering mixed numbers appears only in Scott-Foresman, while estimating the fractional parts of pictorial models appears only in Addison-Wesley.

The researchers also found that, even for many topics covered in all three of the texts, emphases vary considerably among the three books—for some topics, the texts vary in the total percentage of exercises involving those topics. And certain topics appear in varying percentages of the chapters in the three books.

In a second study (6), researchers compared the content of the same three fourth-grade textbooks with the content of several widely used standardized tests: (1) Metropolitan Achievement Tests: Elementary Level, 1978; (2) Stanford Achievement Test (SAT): Intermediate—Level I, 1973; (3) Comprehensive Tests of Basic Skills: Levels I and II, 1976; (4) Iowa Test of Basic Skills: Level 10, 1978. After measuring how well the tests and texts matched on their respective items, the study concluded that no one test is equally well suited for all of the textbooks. As a matter of fact, some striking mismatches were detected. The Addison-Wesley text, for example, covers less than 50 percent of the items on the SAT. Even more striking is the fact that *none* of the texts covers as much as 75 percent of the material on any one of the tests. One implication is clear: don't be too quick to assume that standardized tests cover what is being taught, or that your text covers what is tested.

Couple with this variability the fact that textbooks are becoming longer and contain longer sets of exercises (9)—in response to the back-to-basics movement—and the magnitude of the challenge to teachers becomes clearer. In the Houghton-Mifflin fourth-grade text, for example, there are approximately 3200 computational exercises that involve multiplication or division of whole numbers (8).

In making decisions about textbook use, individual judgment is as important as ever. However, research suggests several areas where teachers must keep a wary eye.

Topics to Heed

All topics demand that teachers do more than merely rely on their texts. There are topics, however, where the need for teacher development and amplification is especially clear.

Counting. As described in the *Research Within Reach* bulletin "Counting Strategies" and in references 14 and 16, children develop a number of counting strategies, especially at the primary level. When a child can use counting to find out how many objects are in a collection and can count out a specified number of objects, that child is using the strategy of *cardinal counting*. Most textbooks use cardinal counting as the basis for approaching addition and subtraction (16) and do not take much heed of the counting strategies developed *after* cardinal counting, such as *counting-back-with-a-tally*, the strategy used when a child confronts the problem, "Bill has 15 cards. How many cards will he have left if he gives Sandra 6 of them?" and reasons, "Let's see, 15 ... 14, 13, 12, 11, 10, 9. Bill will have 9 cards." Because children can use a variety of counting strategies to solve mathematical problems, it is very important for teachers to inject into their classrooms the flexibility that is missing from most textbooks and to identify and welcome each child's own strategies for counting.

Another aspect of number concepts where teachers need to be ready to supplement their texts is *ordinal numbering* (that is, objects ordered in sequence—first, second, third, and so on). In interpreting text pictures where ordinal numbering is represented, some children can mistakenly label positions in reverse order—for example, from right to left, instead of left to right (13).

Money. In a recent study of mathematical thinking in beginning first graders, Hendrickson (7) found a number of topics where young children show more potential than we usually ascribe to them at that age—for example, counting and the solving of addition and subtraction word problems. He also noted one area where texts might *assume* more understanding than really exists among incoming first graders: money relationships, in particular, penny/nickel relationships. Most of the children in the study did not know how many pennies are equivalent to a nickel.

Later in their careers, children may still have difficulty in interpreting money problems that are represented pictorially. For example, Poage and Poage (13) report that, in more than 75 percent of all pictures where coins are illustrated in third-grade textbooks, the heads of the coins are shown. They suggest that this fact might explain the occasionally low performance of children on test problems where only the tail sides of coins are pictured. One implication for teachers: use real coins in classroom discussions before involving students in interpreting pictures of coins.

Volume. Another concept for which textbook coverage is insufficient is volume. In both National Assessments of Educational Progress (NAEP) in mathematics, students were shown a picture of a rectangular solid cut into cubes and were asked to find the number of cubes contained in the solid (2, 3, 11).

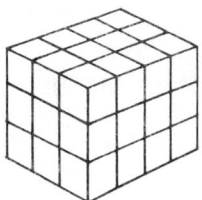

Of course, not all the cubes were visible in the 2-dimensional representation, and the students had to count or compute accordingly. Less than 10 percent of the nine year olds, less than 25 percent of the thirteen year olds, and less than 50 percent of the seventeen year olds, in *each* assessment, answered the question correctly. The message for teachers: develop the concept of volume with concrete experiences *before* resorting to textbook pictures.

The evidence cited above pertains to the specific measurement topics of money and volume. No measurement topic, however, should be taught without regular student involvement in measurement activities. Further information about measurement activities—for time, area, linear measurement, capacity, and so on—can be found in the *Research Within Reach* bulletin "Measurement in Elementary School Mathematics" and the references it recommends.

Estimation. Research has made clear that estimation is a skill that should be woven into the entire elementary school mathematics curriculum. (See the *Research Within Reach* bulletin "Estimation and Mental Arithmetic.") However, current textbooks, as a rule, do a very unsatisfactory job of keeping estimation close to the foreground, where it can be used as a regular strategy in computational work, in problem solving, and in measurement. In the study of the three popular fourth-grade textbooks cited above, estimation appeared in less than 3 percent of the lessons in all three books, and in less than 1 percent of the lessons in two of the books (8).

The lack of attention in textbooks to estimation is unwise and unfortunate. Whether they have the guidance of textbooks or not, all elementary school teachers must make their students more aware of the advantages of using estimation and must help them to develop estimating skills ("How can you tell 287 + 261 is larger than 500? Because 250 + 250 = 500 and both 287 and 261 are larger than 250.").

Problem solving. A survey of five sixth-grade textbooks published between 1973 and 1978 revealed that the total number of verbal problems varied from less than 200 to over 700 (17). When these numbers are considered in light of the evidence that the more problems children solve, the more proficient they become at problem solving (17), it becomes clear that teachers cannot rely totally on their texts for the building of problem-solving skills.

Beyond the issue of the quantity of problems provided by texts, there is the issue of quality. In the recent NAEP assessment, thirteen year olds and seventeen year olds fared badly on a problem that contained extraneous information (11).

> "One rabbit eats 2 pounds of food each week. There are 52 weeks in a year. How much food will 5 rabbits eat in one week?"

Problems like these appear rarely in textbooks, yet the critical skills required are invaluable in the real world.

In a statement about textbook problems, the interpreters of NAEP results have written that "textbooks often promote a mechanistic approach to problem solving, encouraging students to look for 'key words' or clues in a problem to decide which operation to use, rather than trying to extract the meaning of a problem situation" (11, p. 43; see also reference 3). There are a number of ways teachers can supplement their texts' work on problem solving. The *Research Within Reach* bulletin, "Mathematical Problem Solving: Not Just a Matter of Words," can serve as a point of departure.

Beyond the Textbook

The above topics constitute only a partial list for teacher attention in the use of textbooks. In fact, for all concepts, children seem to pass to an appreciation of pictorial and symbolic representations only after they have come to an appreciation of the concrete representation of those concepts (see the *Research Within Reach* bulletin "The Bridge from Concrete to Abstract"). Thus, for example, teachers would be well advised not to rely on *pictures* in their development of the concept of place-value, until they are assured that the child viewing the pictures can manipulate concrete models successfully *and* can make the appropriate connections between the concrete models and the pictures.

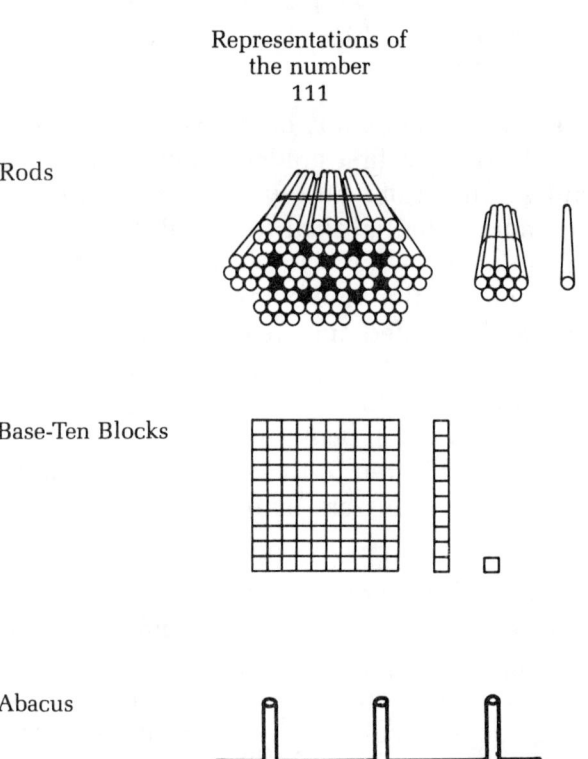

Representations of the number 111

Rods

Base-Ten Blocks

Abacus

When a picture represents the joining of two sets of objects, such as 5 apples with 4 apples, can most children in the intermediate grades give the appropriate number sentence, 5 + 4 = 9? According to data from the recent NAEP assessment in mathematics, this is a reasonable expectation, with approximately 75 percent of nine year olds and 90 percent of thirteen year olds able to translate the picture into the appropriate number sentence.

When the same number sentence was represented on a number line, however, only 25 percent of the nine year olds and fewer than 50 percent of the thirteen year olds taking the test were able to make the correct translation from diagram to number sentence (3, 12).

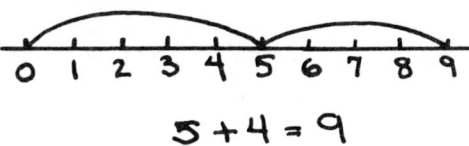

Teachers can help in the transition to picture-reading skills. One recommended approach (1) puts the development of the skills on three successive levels: (1) enumeration—the teacher asks the students to list the objects in a particular picture; (2) description—the teacher gets the students to describe the quality or action depicted; and (3) interpretation—the teacher guides the students in perceiving relationships and making inferences.

Conclusion

There is no doubt that textbooks have a profound influence on student learning. Different texts appear to effect different results in mathematical achievement (15). However, the influence of teacher behavior, while not easy to measure, is also profound. That message runs consistently through these *Research Within Reach* bulletins and is especially strong in "Motivation in Mathematics" and "Meaning in Elementary School Mathematics."

It is the teacher, not the textbook, who can best control student motivation and self-concept, who can cultivate the development of logical reasoning through the interaction of students with peers, and who can capitalize on everyday situations and questions to nurture the growth of problem solving skills. In the hands of an adept teacher, a mathematics textbook can be a vehicle for delivering all those learning experiences, but by itself it is an unsatisfactory, and sometimes unreliable, guide.

References

1. Carpenter, H.M. "Study Skills: Developing Picture-Reading Skills." *The Instructor*, 74 (1964): 37-38; 130.

2. Carpenter, Thomas P., et al. *Results from the First Mathematics Assessment of Educational Progress*. Reston, Virginia: The National Council of Teachers of Mathematics (NCTM), 1978.

* 3. Carpenter, Thomas P., et al. "Results and Implications of the Second NAEP Mathematics Assessment: Elementary School." *Arithmetic Teacher*, 27 (April 1980): 10-14, ff.

4. Educational Products Information Exchange (EPIE). "Research Findings: More About NSAIM." *EPIEgram*, 5, 2 (October 15, 1976).

* 5. Fey, James T. "Mathematics Teaching Today: Perspectives from Three National Surveys." *Arithmetic Teacher*, 27 (October 1979): 10-16.

6. Freeman, Donald; Therese Kuhs; Gabriella Belli; Bob Floden; Lucy Knappen; Andy Porter; Bill Schmidt; and John Schwille. "The Fourth Grade Mathematics Curriculum as Inferred from Textbooks and Tests." Paper presented at Annual Meeting of the American Educational Research Association, Boston, Massachusetts, 1980.

* 7. Hendrickson, A. Dean. "An Inventory of Mathematical Thinking Done by Incoming First-Grade Children." *Journal for Research in Mathematics Education*, 10 (January 1979): 7-23.

8. Kuhs, Therese M., and Donald J. Freeman. *The Potential Influence of Textbooks on Teachers' Selection of Content for Elementary School Mathematics*. Research Series No. 48. East Lansing, Michigan: The Institute for Research on Teaching, 1979.

* 9. McKillip, William D., and Cherie Adler Aviv. "How to Use, Not Abuse, Those Practice Exercises." *Arithmetic Teacher*, 26 (April 1979): 10-13.

References (continued)

*10. National Advisory Committee on Mathematical Education (NACOME). *Overview and Analysis of School Mathematics*. Washington, D.C.: Conference Board of the Mathematical Sciences, 1975.

11. National Assessment of Educational Progress. *Mathematical Applications*, Report No. 09-MA-03. Denver, Colorado: Education Commission of the States, 1979.

12. National Assessment of Educational Progress. *Mathematical Understanding*, Report No. 09-MA-04. Denver, Colorado: Education Commission of the States, 1979.

13. Poage, Melvin, and Esther G. Poage. "Is One Picture Worth One Thousand Words?" *Arithmetic Teacher*, 24 (May 1977): 408-414.

14. Steffe, Leslie P., and Patrick W. Thompson. "Children's Counting in Arithmetical Problem Solving." Paper presented at the Wingspread Conference on the Initial Learning of Addition and Subtraction, Racine, Wisconsin, 1979.

15. Suydam, Marilyn N., and Alan Osborne. *The Status of Pre-College Science, Mathematics, and Social Science Education: 1955-1975. Volume II. Mathematics Education*. Columbus, Ohio: The Information Reference Center for Science, Mathematics, and Environmental Education, 1977.

16. Thompson, Patrick W., and Alba Gonzalez Thompson. "To Count or Not to Count, Is That the Question?" Unpublished manuscript. The University of Georgia, 1978.

17. Zweng, Marilyn J. "The Problem of Solving Story Problems." *Arithmetic Teacher*, 27 (September 1979): 2-3.

---

*The references with asterisks are readings we believe would be especially interesting to and useful for teachers.

Indexes

Subject Index

Abstraction
    Counting Strategies, pp. 18, 19
    From Concrete to Abstract*
    Manipulatives, pp. 21, 27
    Mathematics in Kindergarten, pp. 8, 9

Algorithms
    Algorithms
    Calculators in the Classroom, p. 113
    Diagnosis, p. 41
    From Concrete to Abstract, p. 14
    Meaning in Mathematics, p. 68
    Remediation, p. 48
    Sequence Curriculum, p. 119

Anxiety
    Algorithms, p. 95
    Individualization, p. 86
    Problem Solving, p. 104

Area
    Diagnosis, p. 42
    Measurement, p. 33

Basic facts
    Algorithms, p. 95
    Counting Strategies, p. 19
    Diagnosis, p. 41
    Drill and Other Topics, pp. 73, 74
    Grouping, p. 80
    Meaning in Mathematics, p. 68
    Remediation, p. 47
    Sequence Curriculum, pp. 119, 120

Basic Skills List of National Council of Supervisors of Mathematics (NCSM)
    Sequence Curriculum, p. 121

Beginning Teacher Evaluation Study (BTES)
    Grouping, p. 81
    Individualization, pp. 85, 87

Calculators
    Algorithms, p. 95
    Calculators in the Classroom
    Drill and Other Topics, pp. 74, 75
    Estimation and Mental Arithmetic, p. 110

*Boldface type indicates that an entire *Research Within Reach* bulletin is devoted to the subject.

    Meaning in Mathematics, p. 68
    Problem Solving, p. 102
    Remediation, p. 49
    Sequence Curriculum, p. 119

Cognitive development
    Algorithms, p. 92
    Beyond the Tests, p. 53
    Counting Strategies, pp. 18, 19
    Diagnosis, p. 40
    From Concrete to Abstract, pp. 11, 12, 13, 14
    Grouping, p. 81
    Individualization, p. 85
    Manipulatives, pp. 21-27
    Mastery Learning, p. 57
    Mathematics in Kindergarten, p. 8
    Measurement, pp. 31, 33
    Problem Solving, p. 104
    Sequence Curriculum, p. 116

Comprehensive School Mathematics Program (CSMP)
    From Concrete to Abstract, p. 14
    Individualization, p. 88

Concrete representations
    Algorithms, p. 92
    From Concrete to Abstract
    Manipulatives, p. 22
    Problem Solving, p. 103
    Remediation, p. 48
    Sequence Curriculum, pp. 118, 119
    The Teacher and the Textbook, p. 126

Counting
    Counting Strategies
    Drill and Other Topics, p. 74
    Manipulatives, p. 25
    Mathematics in Kindergarten, p. 8
    The Teacher and the Textbook, p. 124

Cuisenaire rods
    Manipulatives, p. 23

Curriculum sequence
    Algorithms, pp. 93, 94, 95, 96
    Sequence Curriculum

Decimals
    Calculators in the Classroom, p. 113
    Sequence Curriculum, p. 119

Developing Mathematical Processes (DMP) program
   Individualization, p. 88
   Measurement, p. 34
Diagnosis
   Counting Strategies, p. 19
   Diagnosis
   From Concrete to Abstract, p. 12
   Mastery Learning, p. 57
   Remediation, p. 46
Discovery learning
   Meaning in Mathematics, p. 70
   Sequence Curriculum, p. 120
Division
   Algorithms, pp. 95, 96
   Calculators in the Classroom, p. 112
   Drill and Other Topics, p. 75
Drill
   Counting Strategies, p. 20
   Drill and Other Topics
   Meaning in Mathematics, pp. 67, 68, 69
Estimation
   Algorithms, p. 95
   Drill and Other Topics, p. 75
   Estimation and Mental Arithmetic
   Manipulatives, p. 26
   Measurement, pp. 30, 33
   Problem Solving, p. 105
   Remediation, p. 48
   Sequence Curriculum, pp. 120, 121
   The Teacher and the Textbook, p. 125
Evaluation
   Evaluation I: Beyond the Tests
   Diagnosis, p. 41
   Evaluation II: Mastery Learning
   Motivation, p. 62
   The Teacher and the Textbook, p. 124
Flexibility
   Diagnosis, p. 40
   Estimation and Mental Arithmetic, p. 108
   Grouping, p. 80
   Individualization, p. 86
   Mathematics in Kindergarten, p. 9
   Problem Solving, pp. 103, 105
   Remediation, p. 46
Fractions
   Algorithms, p. 94
   Estimation and Mental Arithmetic, p. 109
   Measurement, p. 33
   Sequence Curriculum, pp. 117, 118, 119

Geometry
   Manipulatives, p. 26
   Sequence Curriculum, p. 120
Grouping
   Grouping
   Individualization, p. 88
   Problem Solving, p. 104
   Remediation, p. 46
Hierarchies of learning
   Sequence Curriculum, p. 117
Homework
   Drill and Other Topics, p. 76
Individual learning styles
   Algorithms, p. 95
   Individualization, p. 85
   Manipulatives, p. 22
Individualizing instruction
   Individualization, p. 87
Kindergarten
   Beyond the Tests, p. 54
   Counting Strategies, p. 19
   From Concrete to Abstract, pp. 12, 13
   Mathematics in Kindergarten
Length
   Mathematics in Kindergarten, p. 8
   Measurement, pp. 31, 33
Manipulatives
   Algorithms, p. 92
   Beyond the Tests, p. 53
   Diagnosis, p. 40
   Drill and Other Topics, pp. 73, 74
   From Concrete to Abstract, p. 13
   Grouping, p. 80
   Individualization, p. 87
   Manipulatives
   Problem Solving, p. 103
   Remediation, p. 46
   Sequence Curriculum, p. 118
Mastery Learning
   Evaluation II: Mastery Learning
Mathematics laboratory
   Individualization, p. 88
Meaningful instruction
   Algorithms, p. 93
   Drill and Other Topics, pp. 73, 75
   Meaning in Mathematics
   Sequence Curriculum, p. 120

Measurement
- Manipulatives, p. 26
- Mathematics in Kindergarten, p. 8
- **Measurement**
- Sequence Curriculum, p. 121

Mental arithmetic
- Algorithms, p. 94
- Drill and Other Topics, p. 75
- **Estimation and Mental Arithmetic**

Metric system
- Measurement, p. 34

Motivation
- Beyond the Tests, p. 52
- Calculators in the Classroom, p. 113
- Diagnosis, p. 40
- **Motivation**

Multiembodiments
- Manipulatives, p. 23

National Advisory Committee on Mathematical Education (NACOME)
- Calculators in the Classroom, p. 114
- Individualization, p. 84

National Assessment of Educational Progress — First Assessment
- Algorithms, pp. 93, 94
- Diagnosis, p. 42
- Estimation and Mental Arithmetic, p. 109
- Measurement, pp. 29, 31, 33, 34

National Assessment of Educational Progress — Second Assessment
- Algorithms, pp. 92, 93, 94, 95
- Calculators in the Classroom, p. 112
- Estimation and Mental Arithmetic, p. 110
- Manipulatives, p. 26
- Measurement, pp. 31, 33, 34
- The Teacher and the Textbook, p. 126

Patterns
- Calculators in the Classroom, p. 113
- Mathematics in Kindergarten, p. 9
- Meaning in Mathematics, p. 68
- Problem Solving, p. 102
- Remediation, p. 48

Peer interaction
- Individualization, p. 85
- Problem Solving, pp. 104, 105

Pictorial representations
- Algorithms, p. 92
- From Concrete to Abstract, p. 13
- Manipulatives, p. 25
- Problem Solving, p. 103
- Remediation, p. 48
- Sequence Curriculum, pp. 118, 119
- The Teacher and the Textbook, p. 126

Place value
- Diagnosis, p. 43
- From Concrete to Abstract, p. 13
- Manipulatives, p. 24
- Mathematics in Kindergarten, p. 9

Practice
- Algorithms, p. 95
- Drill and Other Topics, p. 76
- Individualization, p. 87
- Meaning in Mathematics, p. 69
- Problem Solving, p. 103

Problem solving
- Algorithms, p. 92
- Calculators in the Classroom, pp. 113, 114
- Estimation and Mental Arithmetic, p. 109
- Grouping, p. 80
- Individualization, p. 86
- Manipulatives, p. 22
- Mathematics in Kindergarten, pp. 9, 10
- **Problem Solving**
- The Teacher and the Textbook, p. 126

Questions
- Beyond the Tests, p. 54
- Diagnosis, p. 42
- Motivation, p. 62
- Problem Solving, pp. 103, 104

Reasonableness
- Algorithms, p. 95
- Estimation and Mental Arithmetic, p. 109
- Meaning in Mathematics, p. 70
- Problem Solving, p. 105
- Sequence Curriculum, p. 121

Remediation
- Calculators in the Classroom, p. 114
- Manipulatives, p. 24
- **Remediation**

Retention
- Meaning in Mathematics, p. 68
- Sequence Curriculum, pp. 119, 120

Review
- Drill and Other Topics, pp. 74, 76
- Grouping, p. 80

Rounding
- Estimation and Mental Arithmetic, pp. 108, 109

134 Subject Index

Self-concept
    Grouping, p. 80
    Motivation, p. 62
    Remediation, p. 47

Self-paced instruction
    Individualization, p. 84
    Mastery Learning, pp. 57, 59

Symbolic representations
    Algorithms, p. 92
    From Concrete to Abstract, p. 13
    Manipulatives, p. 23
    Remediation, p. 48
    Sequence Curriculum, pp. 118, 119
    The Teacher and the Textbook, p. 126

Systematic errors
    Algorithms, p. 95
    Diagnosis, p. 41
    Manipulatives, p. 24

Teacher expectations
    Motivation, p. 63

Teacher praise
    Drill and Other Topics, pp. 73, 74
    Motivation, pp. 62, 63
    Remediation, p. 47

Teacher-student communication
    Beyond the Tests, p. 54
    Diagnosis, pp. 40, 42
    From Concrete to Abstract, p. 13

    Motivation, p. 62
    Problem Solving, pp. 103, 104, 105
    Remediation, pp. 46, 47

Textbooks
    Manipulatives, p. 26
    Mathematics in Kindergarten, p. 7
    Problem Solving, pp. 101, 102, 104
    The Teacher and the Textbook

Thinking strategies
    Counting Strategies, p. 19
    Drill and Other Topics, p. 74
    Estimation and Mental Arithmetic, p. 108
    Meaning in Mathematics, p. 68
    Sequence Curriculum, p. 120

Time-on-task
    Grouping, p. 81
    Individualization, p. 87
    Mastery Learning, p. 58

Units of measurement
    Manipulatives, pp. 26, 27
    Measurement, pp. 29, 30, 33

Volume
    Manipulatives, p. 26
    Measurement, pp. 29, 33
    The Teacher and the Textbook, p. 125

Whole-class instruction
    Grouping, p. 82

Author Index

Aidala, Gregory
    Calculators in the Classroom, p. 115
Ashlock, Robert B.
    Algorithms, pp. 93, 94, 96, 97
    Diagnosis, pp. 39, 40, 42
    Drill and Other Topics, p. 75
    Estimation and Mental Arithmetic, pp. 107, 109
    Grouping, p. 81
    Manipulatives, p. 24
    Remediation, pp. 46, 47, 48
Austin, Joe Dan
    Drill and Other Topics, pp. 73, 76
Aviv, Cherie Adler
    Drill and Other Topics, pp. 74, 75, 76
    The Teacher and the Textbook, p. 124
Backman, Carl A.
    Diagnosis, p. 41
    Drill and Other Topics, p. 74
Bailey, Terry G.
    Measurement, pp. 30, 33
Bana, Jack
    Beyond the Tests, p. 53
    Manipulatives, p. 23
Barszcz, Edward L.
    Algorithms, p. 96
Beardslee, Edward C.
    Algorithms, pp. 95, 97
    Calculators in the Classroom, p. 113
    Drill and Other Topics, p. 74
Begle, E.G.
    Grouping, pp. 79, 80
Bell, Max S.
    Calculators in the Classroom, pp. 113, 114
Belli, Gabriella
    The Teacher and the Textbook, p. 124
Bereiter, Carl
    Counting Strategies, p. 20
Bierden, James E.
    Grouping, p. 80
Block, James H.
    Mastery Learning, pp. 52, 53

Bloom, Benjamin S.
    Mastery Learning, pp. 51, 53
Borich, Gary D.
    Motivation, pp. 62, 63
Brandau, Linda
    Algorithms, p. 93
Brassell, Anne
    Grouping, p. 80
Braun, Carl
    Motivation, pp. 62, 63
Bright, George W.
    Estimation and Mental Arithmetic, p. 109
    Manipulatives, p. 26
    Measurement, pp. 33, 34
Brophy, Jere E.
    Beyond the Tests, p. 54
    Motivation, pp. 62, 63
    Remediation, p. 47
Brown, Stephen I.
    Grouping, p. 82
Bruner, Jerome
    Manipulatives, p. 21
    Sequence Curriculum, p. 116
Brusch, L.R.
    Mathematics in Kindergarten, pp. 7, 8
Burrows, Charles K.
    Mastery Learning, p. 54
Callahan, Leroy G.
    Drill and Other Topics, p. 75
    From Concrete to Abstract, p. 14
    Meaning in Mathematics, p. 69
    Sequence Curriculum, p. 120
Campbell, N. Jo
    Sequence Curriculum, p. 120
Capps, L.R.
    Algorithms, p. 94
Caravella, Joseph R.
    Calculators in the Classroom, pp. 113, 114
Carpenter, H.M.
    The Teacher and the Textbook, p. 127

Carpenter, Thomas P.
    Algorithms, pp. 92, 93, 94, 96
    Counting Strategies, pp. 17, 18, 19
    Diagnosis, p. 42
    Estimation and Mental Arithmetic, p. 109
    Individualization, p. 88
    Manipulatives, pp. 22, 26, 27
    Mathematics in Kindergarten, p. 10
    Measurement, pp. 30, 31, 32, 33, 34
    Sequence Curriculum, p. 119
    The Teacher and the Textbook, pp. 125, 127
Castaneda, Alberta M.
    Beyond the Tests, p. 54
    Drill and Other Topics, p. 74
    Mastery Learning, p. 58
    Mathematics in Kindergarten, p. 10
Cathcart, W. George
    Grouping, p. 81
    Individualization, pp. 86, 87, 88
    Manipulatives, p. 27
CEMREL, Inc.
    From Concrete to Abstract, p. 14
    Problem Solving, p. 102
Channel, Dwayne E.
    Drill and Other Topics, p. 74
Cifarelli, Victor V.
    Drill and Other Topics, p. 74
Conference Board of the Mathematical Sciences
    Calculators in the Classroom, p. 114
Cox, Linda S.
    Algorithms, p. 96
    Diagnosis, p. 41
Crosswhite, F. Joe
    Grouping, p. 83
    Individualization, p. 87
    Mastery Learning, p. 60
Davis, Edward J.
    Counting Strategies, p. 20
    Drill and Other Topics, pp. 73, 74
    Motivation, p. 62
    Remediation, p. 47
Dawes, Cynthia
    Mathematics in Kindergarten, p. 9
Denmark, Tom
    Diagnosis, p. 40
Dessart, Donald J.
    Algorithms, pp. 93, 94, 95, 96
    Estimation and Mental Arithmetic, p. 109

Meaning in Mathematics, p. 68
    Sequence Curriculum, p. 119
Didactics and Mathematics
    Algorithms, p. 94
    Beyond the Tests, p. 52
    Diagnosis, p. 41
    Drill and Other Topics, p. 76
    Individualization, pp. 85, 87, 88
    Meaning in Mathematics, pp. 68, 70
    Sequence Curriculum, p. 120
Dienes, Zoltan P.
    Manipulatives, pp. 21, 23
Downes, John P.
    Measurement, p. 35
    Sequence Curriculum, p. 120
Dunn, K.
    Grouping, p. 81
    Individualization, pp. 86, 87
Dunn, R.D.
    Grouping, p. 81
    Individualization, p. 86
Easley, Jack
    Algorithms, p. 93
Educational Products Information Exchange
    The Teacher and the Textbook, p. 124
Ellerbruch, Larry W.
    Algorithms, p. 94
    Sequence Curriculum, pp. 118, 119
Engelhardt, Jon M.
    Diagnosis, p. 43
    Motivation, p. 63
    Remediation, p. 48
Erlwanger, Stanley H.
    Mastery Learning, p. 58
Farwest Laboratory for Educational Research and Development
    Drill and Other Topics, p. 76
    Grouping, p. 81
    Individualization, p. 87
Fennema, Elizabeth
    Manipulatives, pp. 23, 24
    Motivation, p. 61
    Problem Solving, p. 103
Fey, James T.
    Grouping, p. 79
    Manipulatives, p. 22
    The Teacher and the Textbook, p. 124

Filby, Nikola
  Individualization, p. 87
Floden, Robert E.
  Beyond the Tests, p. 53
  The Teacher and the Textbook, p. 124
Flournoy, Mary Frances
  Estimation and Mental Arithmetic, p. 109
Folsom, Mary
  Drill and Other Topics, p. 74
  Grouping, p. 80
Fox, Lynn H.
  Motivation, p. 61
Freeman, Donald J.
  Beyond the Tests, p. 53
  The Teacher and the Textbook, pp. 124, 125
Gay, Lorraine R.
  Drill and Other Topics, p. 76
  Individualization, p. 87
  Meaning in Mathematics, p. 69
Gentile, J. Ronald
  Algorithms, p. 96
Gibb, E. Glenadine
  Algorithms, p. 96
  Beyond the Tests, p. 54
  Drill and Other Topics, p. 74
  Mastery Learning, p. 58
  Mathematics in Kindergarten, p. 10
Ginsburg, Herbert
  Beyond the Tests, p. 54
  Calculators in the Classroom, p. 113
  Counting Strategies, pp. 18, 19
  Estimation and Mental Arithmetic, p. 107
  From Concrete to Abstract, p. 13
  Mathematics in Kindergarten, pp. 7, 8, 9
Glennon, Vincent J.
  Drill and Other Topics, p. 75
  From Concrete to Abstract, p. 14
  Meaning in Mathematics, p. 69
  Sequence Curriculum, p. 120
Goldbecker, Sheralyn S.
  Measurement, p. 34
Gonzalez, Alba
  Counting Strategies, p. 18
  The Teacher and the Textbook, p. 124
Good, Thomas L.
  Beyond the Tests, p. 54
  Drill and Other Topics, pp. 74, 75
  Grouping, p. 82
  Meaning in Mathematics, p. 69
  Motivation, pp. 62, 63
  Remediation, p. 47
Gow, Doris T.
  Diagnosis, p. 41
Graeber, Anna O.
  Calculators in the Classroom, p. 113
Green, Geraldine Ann
  Algorithms, p. 94
Grouws, Douglas A.
  Drill and Other Topics, pp. 74, 75
  Grouping, p. 82
  Meaning in Mathematics, p. 69
Hazekamp, Donald W.
  Algorithms, p. 92
  Drill and Other Topics, p. 74
  Meaning in Mathematics, p. 68
  Sequence Curriculum, p. 119
Heddens, James W.
  Drill and Other Topics, p. 77
Hendrickson, A. Dean
  Beyond the Tests, p. 54
  Counting Strategies, pp. 17, 19
  Manipulatives, pp. 22, 25
  The Teacher and the Textbook, p. 125
Herold, Persis Joan
  Diagnosis, p. 41
Hiebert, James
  Algorithms, p. 93
  Counting Strategies, pp. 17, 18, 19
  Individualization, p. 88
  Manipulatives, p. 22
  Mathematics in Kindergarten, p. 10
Higgins, Jon L.
  From Concrete to Abstract, p. 13
  Individualization, p. 87
  Manipulatives, pp. 21, 22, 23, 24, 25
  Measurement, p. 34
  Remediation, p. 48
  Sequence Curriculum, p. 119
Hirstein, James J.
  Measurement, pp. 30, 32, 33
Holtan, Boyd D.
  Problem Solving, p. 104
Hopkins, Martha H.
  Individualization, p. 86

Horwitz, Stephen
    Drill and Other Topics, p. 76
    Meaning in Mathematics, p. 69
Hutchings, Barton
    Algorithms, p. 97
    Remediation, p. 48
Hynes, Michael C.
    Algorithms, pp. 97, 98
    Manipulatives, pp. 24, 25
Immerzeel, George
    Calculators in the Classroom, p. 113
    Problem Solving, p. 102
    Remediation, p. 49
Inskeep, James E., Jr.
    Diagnosis, p. 41
    Measurement, pp. 30, 31, 34
    Remediation, p. 47
Jackson, Robert L.
    Manipulatives, p. 26
Jensen, Rosalie
    Sequence Curriculum, p. 120
Johnson, David C.
    Estimation and Mental Arithmetic, p. 109
Johnson, Hiram D.
    Sequence Curriculum, p. 120
Johnson, Stanley W.
    Beyond the Tests, p. 53
    Diagnosis, p. 41
Junge, Charlotte W.
    Algorithms, p. 97
    Diagnosis, p. 43
    Grouping, p. 81
    Individualization, pp. 87, 88
    Manipulatives, p. 24
    Remediation, pp. 46, 47, 48
Kamii, Constance K.
    Individualization, p. 85
Kane, Robert B.
    Sequence Curriculum, pp. 117, 119
Kasnic, Michael J.
    Calculators in the Classroom, p. 114
Kersh, Mildred E.
    Mastery Learning, p. 59
Kilpatrick, Jeremy
    Problem Solving, p. 106
Kirkpatrick, Joan
    Mathematics in Kindergarten, p. 10
    Problem Solving, p. 106
Knappen, Lucy
    The Teacher and the Textbook, p. 124
Knifong, J. Dan
    Problem Solving, p. 104
Kramer, Klaas
    Estimation and Mental Arithmetic, p. 111
Krulik, S.
    Grouping, p. 80
    Problem Solving, p. 106
Krutetskii, V.A.
    Problem Solving, p. 106
Kuhs, Therese
    The Teacher and the Textbook, pp. 124, 125
Lankford, Francis G., Jr.
    Algorithms, p. 96
    Diagnosis, p. 42
LeBlanc, John F.
    Problem Solving, pp. 101, 102, 105
Lester, Frank K., Jr.
    Problem Solving, p. 103
Leutzinger, Larry P.
    Counting Strategies, p. 20
    Drill and Other Topics, p. 74
    Manipulatives, p. 25
Lewis, Ruth
    Manipulatives, pp. 26, 27
    Measurement, p. 33
Liedtke, Werner
    Individualization, p. 86
Lovell, Kenneth
    From Concrete to Abstract, pp. 11, 12, 13
Mahaffey, Michael I.
    Beyond the Tests, p. 55
Mason, Marguerite
    Calculators in the Classroom, p. 113
McKillip, William D.
    Drill and Other Topics, pp. 74, 75, 76
    Mathematics in Kindergarten, p. 9
    Problem Solving, p. 102
    The Teacher and the Textbook, p. 124
Merseth, Katherine Klippert
    Algorithms, p. 92
    From Concrete to Abstract, p. 13

Mathematics in Kindergarten, p. 9
Meaning in Mathematics, p. 68
Remediation, p. 48

Meyerson, Lawrence N.
Grouping, p. 82

Moser, James M.
Algorithms, p. 93
Calculators in the Classroom, p. 113
Counting Strategies, pp. 17, 18, 19
Individualization, p. 88
Manipulatives, p. 22
Mathematics in Kindergarten, p. 10

National Advisory Committee on Mathematical Education
Individualization, pp. 84, 85
The Teacher and the Textbook, p. 128

National Assessment of Educational Progress
Algorithms, pp. 92, 95
Calculators in the Classroom, p. 112
Drill and Other Topics, pp. 75, 76
Estimation and Mental Arithmetic, p. 110
Manipulatives, p. 26
Measurement, pp. 31, 32, 33
The Teacher and the Textbook, pp. 125, 126, 127

National Council of Teachers of Mathematics
Measurement, pp. 34, 35

Nelson, Glenn
Counting Strategies, p. 20
Drill and Other Topics, p. 74
Manipulatives, p. 25

Nelson, L. Doyal
Beyond the Tests, p. 53
Manipulatives, pp. 23, 28
Mathematics in Kindergarten, p. 10
Measurement, p. 35
Problem Solving, p. 106

Ockenga, Earl
Calculators in the Classroom, p. 113
Problem Solving, p. 102
Remediation, p. 49

O'Daffer, Phares
Estimation and Mental Arithmetic, pp. 108, 109

Okey, James R.
Mastery Learning, pp. 52, 54

Osborne, Alan
The Teacher and the Textbook, pp. 123, 124, 127

Pagni, David L.
Measurement, p. 31

Payne, Joseph N.
Algorithms, p. 94
Diagnosis, p. 40
Estimation and Mental Arithmetic, p. 108
From Concrete to Abstract, pp. 13, 14
Mathematics in Kindergarten, p. 9
Sequence Curriculum, pp. 117, 118, 119

Pearson, James R.
Motivation, p. 62

Phillips, E. Ray
Sequence Curriculum, pp. 117, 119

Pigge, F.
Meaning in Mathematics, p. 69

Pikaart, Len
Remediation, p. 43

Poage, Esther G.
The Teacher and the Textbook, p. 125

Poage, Melvin
The Teacher and the Textbook, p. 125

Porter, Andrew C.
Beyond the Tests, p. 53
The Teacher and the Textbook, p. 124

Prigge, Glenn R.
Individualization, p. 87
Manipulatives, p. 26

Rathmell, Edward C.
Counting Strategies, p. 19
Drill and Other Topics, p. 74
Estimation and Mental Arithmetic, p. 108
From Concrete to Abstract, pp. 13, 14
Mathematics in Kindergarten, p. 9
Meaning in Mathematics, p. 68
Sequence Curriculum, p. 120

Rea, Robert E.
Counting Strategies, p. 19
Mathematics in Kindergarten, p. 8
Measurement, p. 30

Reisman, Fredricka K.
Beyond the Tests, p. 53
Diagnosis, pp. 41, 42
Remediation, p. 48

Reys, Robert E.
Algorithms, p. 92
Calculators in the Classroom, pp. 113, 114
Counting Strategies, p. 19

Grouping, p. 81
Individualization, p. 87
Manipulatives, pp. 23, 26, 28
Mastery Learning, p. 60
Mathematics in Kindergarten, p. 8
Meaning in Mathematics, p. 68
Measurement, p. 30
Problem Solving, p. 102

Riedesel, C. Alan
Diagnosis, p. 40
Mathematics in Kindergarten, pp. 7, 8

Rim, Eui-Do
Calculators in the Classroom, p. 113

Rising, Gerald R.
Grouping, p. 82

Roberts, Gerhard H.
Algorithms, p. 98

Robinson, G. Edith
Beyond the Tests, p. 55
Manipulatives, pp. 23, 26
Measurement, pp. 29, 30, 31, 34
Sequence Curriculum, p. 120

Romberg, Thomas A.
Motivation, p. 62

Schall, William E.
Estimation and Mental Arithmetic, pp. 109, 110

Schmidt, William H.
Beyond the Tests, p. 53
The Teacher and the Textbook, p. 124

Schneider, E. Joseph
Grouping, p. 81
Individualization, pp. 85, 87

Schoen, Harold L.
Individualization, p. 85
Mastery Learning, p. 59
Sequence Curriculum, p. 120

Schulz, Richard W.
Remediation, pp. 45, 47

Schwille, John
The Teacher and the Textbook, p. 124

Scott, A.
Algorithms, p. 96

Sharma, Mahesh
Individualization, p. 87
Manipulatives, p. 23

Sherman, Julia
Motivation, p. 61

Shumway, Richard
From Concrete to Abstract, p. 14

Shuster, A.H.
Meaning in Mathematics, p. 69

Smith, Jeffrey K.
Mastery Learning, p. 57

Smith, Seaton E., Jr.
Drill and Other Topics, p. 74

Speer, William R.
Individualization, p. 86

Steffe, Leslie P.
Counting Strategies, pp. 18, 19
Drill and Other Topics, p. 74
Measurement, p. 30
The Teacher and the Textbook, p. 124

Stevenson, Harold W.
Beyond the Tests, pp. 53, 54
From Concrete to Abstract, pp. 12, 13
Manipulatives, p. 23
Problem Solving, p. 104

Suydam, Marilyn N.
Algorithms, pp. 92, 93, 94, 95, 96
Beyond the Tests, pp. 52, 53
Calculators in the Classroom, pp. 112, 113
Diagnosis, p. 43
Drill and Other Topics, p. 74
Estimation and Mental Arithmetic, p. 109
From Concrete to Abstract, p. 13
Grouping, pp. 79, 81
Individualization, p. 89
Manipulatives, pp. 21, 22, 23, 24, 25
Mathematics in Kindergarten, p. 8
Meaning in Mathematics, pp. 67, 68, 69
Motivation, p. 62
Problem Solving, pp. 102, 103, 104
Remediation, p. 48
Sequence Curriculum, pp. 119, 121
The Teacher and the Textbook, pp. 123, 124, 127

Thompson, Patrick W.
Counting Strategies, pp. 18, 19
The Teacher and the Textbook, p. 124

Thornton, Carol A.
Counting Strategies, p. 19
Drill and Other Topics, p. 74
Estimation and Mental Arithmetic, p. 108
Meaning in Mathematics, p. 68
Problem Solving, p. 106
Sequence Curriculum, p. 120

Trafton, Paul R.
    Drill and Other Topics, p. 74
    Estimation and Mental Arithmetic, pp. 108, 109
    Meaning in Mathematics, p. 70
    Sequence Curriculum, p. 120

Trimmer, Ronald G.
    Problem Solving, p. 103

Trown, Anne
    Individualization, pp. 86, 88

Underhill, Robert G.
    Diagnosis, p. 41

Unks, Nancy J.
    Calculators in the Classroom, p. 113

Van Engen, Henry
    Algorithms, p. 96

Washbon, Carolynn A.
    Drill and Other Topics, p. 75

Weaver, J. Fred
    Diagnosis, p. 44
    Grouping, p. 79
    Manipulatives, pp. 23, 24
    Mathematics in Kindergarten, p. 8
    Meaning in Mathematics, pp. 67, 68, 69
    Motivation, p. 62
    Problem Solving, pp. 102, 103, 104
    Sequence Curriculum, p. 122

Wheatley, Charlotte L.
    Algorithms, pp. 95, 96
    Calculators in the Classroom, pp. 113, 114
    Problem Solving, p. 103

Wheatley, Grayson
    Algorithms, pp. 95, 96, 97
    Calculators in the Classroom, p. 113
    Counting Strategies, p. 17
    Drill and Other Topics, pp. 74, 76
    From Concrete to Abstract, p. 14

Wilson, James W.
    Remediation, p. 43

Wirtz, Robert
    Mathematics in Kindergarten, p. 9

Wisconsin R&D Center for Individualized Learning
    Individualization, p. 88
    Measurement, p. 34

Woodward, Ernest
    Individualization, p. 88

Worthen, Blaine R.
    Meaning in Mathematics, p. 70

Zepp, Raymond
    Algorithms, p. 95
    Estimation and Mental Arithmetic, p. 109

Zweng, Marilyn J.
    Problem Solving, p. 104
    The Teacher and the Textbook, p. 126